Charge Distributions
and Chemical Effects

Sándor Fliszár

Charge Distributions and Chemical Effects

A New Approach to the
Electronic Structure and Energy
of Molecules

Springer-Verlag
New York Berlin Heidelberg Tokyo

Sándor Fliszár
Département de chimie
Faculté des arts et des sciences
Université de Montréal
Case postale 6210, Succursale A
Montréal, Québec, Canada H3C 3Vl

With 15 Illustrations

Library of Congress Cataloging in Publication Data
Fliszár, Sándor.
  Charge distributions and chemical effects.
  Bibliography: p.
  Includes index.
  1. Molecular structure.   I. Title.
QD461.F54 1983   541.2′2   83-6682

Typeset by Asco Trade Typesetting Limited, Hong Kong.
Printed and bound by R.R. Donnelley & Sons, Harrisonburg, Virginia.
Printed in the United States of America.

9 8 7 6 5 4 3 2 1

ISBN 0-387-90854-4   Springer-Verlag   New York Berlin Heidelberg Tokyo
ISBN 3-540-90854-4   Springer-Verlag   Berlin Heidelberg New York Tokyo

To the memory of my father

Preface

The energy of a molecule can be studied with the help of quantum theory, a satisfactory approach because it involves only basic and clearly identified physical concepts. In an entirely different approach, the molecular energy can be broken down into individual contributions reflecting chemical bonds plus a host of subsidiary "effects", like γ-*gauche*, *skew* pentane, ring-strain, etc., giving an overall picture in terms of topological characteristics. The latter approach can be successful, particularly if a sufficient number of particular topological situations have been parametrized (which is an empirical way of "understanding" chemistry), but also contains the seed for difficulties. Indeed, the danger exists of unduly ascribing a physical meaning to corrective terms whose function is primarily to account in an empirical fashion for discrepancies between "expected" and observed results. The link between this type of empirical approach and the knowledge that the ground state energy is uniquely determined by the electron density is lost somewhere along the road, although some of the "steric effects" are here and there vaguely traced back to electronic effects.

The approach presented in this monograph goes back to the fundamentals in that it is exclusively based on interactions involving nuclear and electronic charges. Confining the study to molecules in their equilibrium geometry, the problem of molecular energies is reduced to its electrostatic aspects, explicitly involving local electron populations. In this manner, it becomes possible to break down the molecular energy into local contributions, namely the individual chemical bonds, featuring the effects of local atomic charges. In short, a description of the molecule is given in terms of charge distributions.

Following a brief crash-exposé on the calculation of atomic charges (Chapter 1), the validity of theoretical charges, as well as their role in familiar problems of inductive effects, are examined in detail (Chapters 2 and 3). Next, convenient ways of deducing adequate sets of charges from ^{13}C and ^{17}O NMR spectra are discussed (Chapter 4), stressing both the usefulness and limitations of charge–shift correlations. Chapter 5 describes the calculation of individual bond energy contributions and of molecular energies in terms of local atomic charges, using only basic physical concepts which are easily identified, namely, nuclear–nuclear and nuclear–electronic interaction energies. The comparison between calculated and experimental energies (Chapters 6–9) reveals an overall agreement which is well within experimental uncertainties. The fine-tuning of the energies due to subtle local electronic properties is shown to account for a number of well-known conformational effects, e.g., the difference between the boat and chair forms of cyclohexane or the axial $vs.$ the equatorial forms of methylcyclohexane. The role of vibrational energies is analyzed in Chapter 7.

This monograph is directed toward both organic and physical chemists. Hopefully, it will be of general interest. Indeed many of my colleagues have found it interesting, for example, that a gain of 0.001 electron on a hydrogen atom stabilizes a CH bond by 0.632 kcal/mol or that the comparison of twistane with its structural isomer adamantane reveals a gain in stability (1.53 kcal/mol) for the 16 CH bonds in twistane and a loss in stability of its carbon skeleton (10.69 kcal/mol) due to lower electron populations of its carbon atoms, these results being entirely deduced from ^{13}C NMR spectra. The size of this monograph is still modest because it represents only a beginning in new areas of research. Hopefully, organic chemists will recognize many familiar aspects and exploit these results by contributing examples. For this reason, parts of this book are presented as descriptive interpretations of the theoretical results and are aimed at a wider audience. This is, at least, what I hope, and so does my publisher.

ACKNOWLEDGMENTS

Much of the work described here has been the result of a collaborative effort with a number of graduate and post-graduate students at the Université de Montréal. These include Hervé Henry, Gérard Kean, Jacques Grignon, Annick Goursot, Aniko Foti, Marie-Thérèse Béraldin, Jacques Bridet, Marielle Foucrault, Jean-Louis Cantara and Guy Cardinal. Helpful discussions with a number of colleagues, namely with Professors R.G. Parr, G. Del Re, V.H. Smith, R.F.W. Bader, P. Politzer, A. Julg, S. Odiot, S. Califano, D. Gravel, E. Vauthier, and D. Salahub, are also acknowledged. Finally, I wish to express my sincere appreciation to Professor Oktay Sinanoğlu for his valuable comments which have been of great help in the preparation of the final manuscript. The efforts of Claire Potvin in preparing the typescript are greatly appreciated. Permission to reproduce material has been granted by the *Canadian Journal of Chemistry*, the *Journal of the American Chemical Society*, the *Journal of Chemical Physics*, and the *Physical Review*.

The support of the National Research Council of Canada and of the "France-Québec Exchange Program" are gratefully acknowledged.

Sándor Fliszár

Contents

CHAPTER 1

Electronic Charge Distributions

1. INTRODUCTION

One of the most popular concepts in chemistry is that of charge distributions in molecules: the vital significance of electron density in the theory of molecular structure is firmly assessed by the Hohenberg–Kohn theorem[1] which shows that under certain assumptions a non-degenerate ground state is a unique functional of the electron density. Unfortunately, only the existence, but not the form, of the functional was determined.

Pictorial presentations of charge densities can be offered in a number of ways. In the familiar contour map type[2], for example, contours corresponding to various values of the charge density (or of its difference with respect to the superposition of the free atoms) are plotted for different points in a specified plane of the molecule[3-7]. This type of presentation is certainly a realistic one. Contour maps alone, however, do not yet tell us how much charge can be assigned in a meaningful way to the individual centers. Appropriate partitioning surfaces, e.g., those defined in Bader's virial partitioning method[8], lead us from a spatial description of electron density to the concept of atomic charge, i.e., to a distribution of the total number of electrons in a system among the atoms involved.

The partitioning of the electron density into atomic charges offers a vivid description which is adequate in the discussion of a number of chemical problems. Indeed, substituent and conformational effects governing local charges and bond polarities are best described in terms of atomic charges. Moreover, these charges are instrumental in calculations of bond and molecular energies. Because of this, a major effort in the study of charge distributions is justified. Our discussion is non-relativistic within the Born–Oppenheimer approximation and involves only the static charge distributions.

Atomic charges are most conveniently calculated from molecular wave functions following Mulliken's population analysis[9]. Results derived therefrom are a solid basis for the discussion of some aspects regarding charge distributions, in spite of strong criticisms regarding the method itself. Modifications of Mulliken's analysis are considered at a later stage, as difficulties begin to surface in attempted applications to physical problems.

2. ELECTRON POPULATION ANALYSIS

The knowledge of the molecular wave function enables us to determine the electron density at any given point in space. Here we inquire more specifically about the amount of electronic charge that can be associated in a meaningful way with each individual atom of an N-electron system. Mulliken's population analysis for extracting atomic charges from wave functions is rooted in the LCAO formulation and is not directly applicable to other types of wave functions. If the ith normalized molecular orbital is written

$$\phi_i = c_{Ai}\phi_A + c_{Bi}\phi_B$$

as a linear combination of normalized atomic orbitals centered on atoms A and B, with real coefficients, and v_i is the occupation number of the ith molecular orbital, Mulliken's analysis associates $v_i c_{Ai}^2$ electrons with atom A and $v_i c_{Bi}^2$ with atom B, in the ith MO. These are the so-called "net" atomic populations. Moreover, another term arising from the space integral $\int(c_{Ai}\phi_A + c_{Bi}\phi_B)^2 d\tau$ must be partitioned among A and B, i.e., the integrated electron density "between" the two nuclei arising from the ith molecular orbital. This quantity is the "overlap population"

$$2v_i c_{Ai} c_{Bi} \langle \phi_A | \phi_B \rangle$$

where $\langle \phi_A | \phi_B \rangle = S_{AB}$ is the overlap integral between the two atomic orbitals. The "gross atomic population" is deduced, in Mulliken's scheme, by assuming that the overlap population can be divided equally between the two atoms. This half-and-half partitioning is, however, problematic, especially in a heteropolar situation. As far as the dipole moment is concerned, this assumption is equivalent to saying that the center of gravity of the overlap population is equal to the midpoint of the vector joining the two nuclei. As a consequence of this approximation, the dipole moment constructed from the two charges, or even the bond polarity ($A^+ B^-$ or $A^- B^+$), is incorrect in most cases. Leaving aside temporarily this kind of criticism, the gross atomic population arising from the ith molecular orbital is given by

$$v_i(c_{Ai}^2 + c_{Ai} c_{Bi} S_{AB})$$

and the total number of electrons associated with atom A is obtained from a summation over all the occupied molecular orbitals.

This analysis is readily generalized to molecular orbitals which have more than one component atomic orbital with a given atomic center. With $c_{r_k i}$ representing now the coefficient of the rth type of atomic orbital ($1s$, $2s$, etc.) of the atom k in the ith molecular orbital, we describe the latter by

$$\phi_i = \sum_{r_k} \chi_{r_k} c_{r_k i}$$

where the summation extends over all the appropriate normalized basis functions χ_{r_k} and the subindex k labels the different nuclei in the system. The corresponding overlap population associated with atoms k and l due to atomic orbitals of type r and s, respectively, is then

$$2 v_i c_{r_k i} c_{s_l i} S_{r_k s_l}.$$

Finally, Mulliken's analysis yields the population, $N(k)$, on atom k from the appropriate sums over all doubly occupied molecular orbitals i and over all types of basis functions, i.e.,

$$N(k) = 2 \sum_i \sum_r \left(c_{r_k i}^2 + \sum_{l \neq k} c_{r_k i} c_{s_l i} S_{r_k s_l} \right) \tag{1.1}$$

The quantity which is referred to in most cases is the "net atomic charge" q_k defined as

$$q_k = Z_k - N(k) \tag{1.2}$$

where Z_k is the nuclear charge of center k. With this definition, q_k is negative when an atom in a molecule carries a number of electrons in excess of that of the neutral atom. This definition is generally valid, independently of the mode of obtaining $N(k)$. When $N(k)$ is derived from a Mulliken population analysis, as in Eq. 1.1, q_k is referred to as Mulliken net atomic charge.

The $N(k)$ result expressed by Eq. 1.1 implies, of course, the half-and-half partitioning of all overlap population terms among the centers k, l, ... involved. The division of the overlap charge has concerned many authors[10,11]. Indeed, while the usual (Mulliken) half-and-half assignment is easy to defend in situations involving partners of equal nature, this may be a less good approximation in cases involving dissimilar atoms. In fact, because of the large weight of the overlap contributions to total atomic populations, even minor (1–3%) departures from the usual half-and-half partitioning result in sizable modifications of the net atomic charges. A clear recognition of this point is essential in any study involving Mulliken-type electron distributions. Assuming now a modified (and more general) mode of distributing overlap populations, one obtains for the population on center k that

$$N(k) = 2 \sum_i \sum_r \left(c_{r_k i}^2 + \sum_{l \neq k} c_{r_k i} c_{s_l i} S_{r_k s_l} \lambda_{r_k s_l} \right) \tag{1.3}$$

where the weighting factor $\lambda_{r_k s_l}$ causes the departure from the usual halving of the overlap terms. Mulliken's charges correspond to $\lambda_{r_k s_l} = 1$. One may employ the coefficients λ in order to conserve the dipole moment, for example, or (more generally) in order to define a scaling of atomic charges satisfying some property, e.g., ^{13}C NMR shifts (Chapter 4) or energies of atomization (Chapter 6). This sort of undertaking leads, ultimately, to an experimental partitioning of overlap populations[11].

In terms of the difference

$$2\sum_i \sum_r \sum_{l \neq k} c_{r_k i} c_{s_l i} S_{r_k s_l}(1 - \lambda_{r_k s_l}) = \sum_{l \neq k} p_{kl} \tag{1.4}$$

between Mulliken charges (Eq. 1.1) and those given by Eq. 1.3, one obtains for the net atomic charge q_k of atom k (Eq. 1.2) that

$$q_k = q_k^{\text{Mulliken}} + \sum_{l \neq k} p_{kl} \tag{1.5}$$

where the charge normalization is ensured by the condition $p_{lk} = -p_{kl}$. Equation 1.5 is instrumental in forthcoming analyses of charge-dependent properties.

This brief account does not render justice to the great many studies which have dealt with the problem of charge partitioning. Mulliken's scheme and its modification (Eq. 1.3) certainly represent the simplest ones in LCAO theory. Since this is sufficient for our intended purpose, we refrain from going into further considerations. These are well reviewed in the literature[7].

3. BASIS FUNCTIONS

Mulliken population analyses are commonly carried out using a variety of LCAO methods, both semi-empirical and *ab initio*. The latter involve the solution, based on the energy variation principle, of the Hartree–Fock equations. The SCF solution is obtained by means of an expansion of the molecular orbitals over a finite set of basis orbitals, where the basis functions are usually analytical atomic type orbitals centered at the atomic sites.

The simplest set in LCAO–SCF calculations consists of the minimal number of functions required for the description of the fundamental atomic orbitals, e.g., $1s$ for hydrogen, $1s$, $2s$, $2p_x$, $2p_y$, $2p_z$ for carbon, etc. In so-called minimal basis set calculations, one Slater-type orbital (STO) of the general form

$$\chi_{ns}(\zeta_n, r) = \left[\frac{\zeta_n^{2n+1} 2^{2n+1}}{\pi(2n)!}\right]^{1/2} r^{n-1} e^{-\zeta_n r}$$

$$\chi_{np}(\zeta_n, r) = 3^{1/2}\chi_{ns}(\zeta_n, r) \times \begin{cases} \cos\theta \\ \sin\theta\cos\phi \\ \sin\theta\sin\phi \end{cases}$$

is used for each occupied atomic orbital in the constituent atoms. Because of the molecular environment, the ζ exponents must be optimized for the molecule in question[12]. Sometimes, recommended "standard" ζ exponents which were optimized for atoms belonging to a given class of molecules (e.g., the C and H atoms in alkanes) are used without reoptimization for other molecules belonging to the same class. Charge results, however, are very sensitive to exponent optimization and the most consistent results are obtained when the ζ optimization is carried out at its best for each atom (or group of equivalent atoms) of the molecule under study[13]. For computational reasons, Slater-type orbitals are conveniently represented by linear combinations of Gaussian primitives. For example, the STO–3G set is a minimal basis set in which each STO is represented by an equivalent linear combination of 3 Gaussians.

In current practice, more sophisticated basis sets are favored. Namely, if we add a second (or more) Slater-type orbital for describing each of the atomic orbitals, we speak of double (or higher zeta) basis sets. Of course, this leads to improved calculations and is a requirement for an adequate description of the charge density in the important regions, namely, in the proximity of the nuclei. This point is important because the variational principle will place the primary emphasis on the regions closer to the nuclei because of the electron–nuclear attraction term in the Hamiltonian operator. Therefore, only large basis sets yield hopefully valid results in calculations of, say, nuclear–electronic potential energies. If, however, only atomic charges of the Mulliken-type are desired, STO–3G calculations turn out to be quite adequate, moreover as with this sort of minimal basis set the all-important extensive optimizations of the ζ exponents remain within reach of computational feasibility—a point which is important because atomic charges are very sensitive to ζ optimization.

It is instructive to examine the influence of the basis set on Mulliken population analyses derived from SCF calculations. This is best illustrated by numerical examples. STO–3G calculations using "standard" ζ exponents indicate that the carbon atom in methane carries a net charge (Eq. 1.2) of -73 me (1 me $= 10^{-3}$ electron); full optimization of the ζ exponents yields[13] -48.92 me. "Better" *ab initio* results, i.e., those derived using larger Gaussian basis sets, indicate, however, much larger charge separations than those derived using a minimal basis set. A consideration of Leroy's $7s3p|3s$ results[14] (q_c in $CH_4 = -790$ me) or Allen's results[15] ($q_c = -1.072$ electron in CH_4) then raises the questions: (*i*) why do alkanes with such positive H atoms never form hydrogen bonds, or (*ii*) why are *ab initio* results for charge distributions less credible for (good) large Gaussian basis sets than those obtained from the less sophisticated 3G calculations?

This sort of result is hardly encouraging. Fortunately, things are not quite as bad as these examples would seem to indicate even though eventually it turns out that the carbon in methane is, in fact, electron deficient, i.e., that the CH polarity is C^+—H^-. The worrisome problems raised here can be resolved with the aid of the modified population analysis (Eqs. 1.3–1.5). Before doing so, however, let us assess whatever positive aspects are contained in conventional charge analyses, by an examination of simple hydrocarbons.

REFERENCES

1. P. Hohenberg and W. Kohn, *Phys. Rev.*, **136B**, 864 (1964); W. Kohn and L.J. Sham, *Phys. Rev.*, **140A**, 1133 (1965).
2. J.R. Van Wazer and I. Absar, "Electron Densities in Molecules and Molecular Orbitals", Academic Press, New York, NY, 1975.
3. M. Roux, S. Besnainou, and R. Daudel, *J. Chim. Phys.*, **53**, 218, 939 (1956); M. Roux, *J. Chim. Phys.*, **55**, 754 (1958); *ibid.*, **57**, 53 (1960); M. Roux, M. Cornille, and L. Burnelle, *J. Chem. Phys.*, **37**, 933 (1972).
4. R.F.W. Bader, in *International Review of Science, Physical Chemistry, Series Two*, Vol. 1, A.O. Buckingham, Ed., Butterworths, London, 1975, pp. 43–78; P.E. Cade, R.F.W. Bader, W.H. Henneker, and I. Keaveny, *J. Chem. Phys.*, **50**, 5313 (1969); R.F.W. Bader and A.D. Bandrauk, *J. Chem. Phys.*, **49**, 1653 (1968); R.F.W. Bader and W.H. Henneker, *J. Am. Chem. Soc.*, **87**, 3063 (1965); *ibid.*, **88**, 280 (1966); R.F.W. Bader, W.H. Henneker, and P.E. Cade, *J. Chem. Phys.*, **46**, 3341 (1967).
5. P. Coppens, in *International Review of Science, Physical Chemistry, Series Two*, Vol. 2, J.M. Robertson, Ed., Butterworths, London, 1975, pp. 21–56; P. Coppens and E.D. Stevens, *Adv. Quantum Chem.*, **10**, 1 (1977); B. Dawson, "Studies of Atomic Charge Density by X-Ray and Neutron Diffraction, A Perspective", Pergamon Press, London, 1975; B.J. Ransil and J.J. Sinai, *J. Chem. Phys.*, **46**, 4050 (1967); F.L. Hirschfeld and S. Rzotkiewicz, *Mol. Phys.*, **27**, 319 (1974); B.J. Ransil and J.J. Sinai, *J. Am. Chem. Soc.*, **94**, 7268 (1972).
6. E.A. Laws and W.N. Lipscomb, *Israel J. Chem.*, **10**, 77 (1972).
7. V.H. Smith, *Phys. Scripta*, **15**, 147 (1977); A. Julg, *Top. Curr. Chem.*, **58**, 1 (1975).
8. R.F.W. Bader, *Acc. Chem. Res.*, **8**, 34 (1975); G.R. Runtz and R.F.W. Bader, *Mol. Phys.*, **30**, 129 (1975); R.F.W. Bader and G.R. Runtz, *Mol. Phys.*, **30**, 117 (1975); R.F.W. Bader and R.R. Messer, *Can. J. Chem.*, **52**, 2268 (1974); S. Srebrenik and R.F.W. Bader, *J. Chem. Phys.*, **63**, 3945 (1975).
9. R.S. Mulliken, *J. Chem. Phys.*, **23**, 1833, 1841, 2338, 2343 (1955).
10. P.-O. Löwdin, *J. Chem. Phys.*, **18**, 365 (1950); *ibid.*, **21**, 374 (1953); E.R. Davidson, *J. Chem. Phys.*, **46**, 3320 (1967); K. Jug. *Theor. Chim. Acta*, **31**, 63 (1973); *ibid.*, **39**, 301 (1975).
11. S. Fliszár, A. Goursot, and H. Dugas, *J. Am. Chem. Soc.*, **96**, 4358 (1974); S. Fliszár, *Can. J. Chem.*, **54**, 2839 (1976); H. Henry and S. Fliszár, *J. Am. Chem. Soc.*, **100**, 3312 (1978); S. Fliszár, *J. Am. Chem. Soc.*, **102**, 6946 (1980); M.-T. Béraldin, E. Vauthier, and S. Fliszár, *Can. J. Chem.*, **60**, 106 (1982).
12. W.J. Hehre, R.F. Stewart, and J.A. Pople, *J. Chem. Phys.*, **56**, 2657 (1969).
13. G. Kean and S. Fliszár, *Can. J. Chem.*, **52**, 2772 (1974).
14. J.M. André, P. Degand, and G. Leroy, *Bull. Soc. Chim. Belg.*, **80**, 585 (1971).
15. J.E. Williams, V. Buss, and L.C. Allen, *J. Am. Chem. Soc.*, **93**, 6867 (1971).

CHAPTER 2

Charge Analysis of Simple Alkanes

1. INTRODUCTION

The success of a quantum mechanical description of a molecule is usually evaluated by its ability to reproduce correctly some observable quantity, such as spectroscopic information, rotational barriers, conformations, etc. However, different quantum mechanical approaches, though each one may be successful in its own right, are known to generate charge distributions which vary depending upon the method. This is disturbing if one tends to consider molecular wave functions and charge densities as part of the description of the molecular reality.

   In this situation, one is thus restricted to express satisfaction with the fact that the different methods of calculation usually reproduce the same *trends* in charge distributions. Such an attitude may be justified, at least in part, on the grounds that in a number of cases the trends predicted by a variety of theoretical methods seem intuitively correct, i.e., that they often agree with what one would expect from chemical evidence. In other words, theoretical trends are accepted when they make sense in terms of the intuition derived from experimental observations. Unfortunately, it is just as easy to produce counter-examples. Indeed, MO calculations indicate the phenyl ring in toluene to be electron poorer than the phenyl group in benzene, which is contrary to common chemical expectation. For example, Hoffmann's extended Hückel calculations[1] indicate a net charge for the C_6H_5 group of -21 me in toluene and -101 me in benzene. Pople's *ab initio* results[2] indicate the same trend for the charge in the C_6H_5 group: -37 me in toluene, and -51 me in benzene. Both Hoffmann's and Pople's results seemingly suggest that hydrogen is a better electron donor than methyl—at odds with the usual interpretation of the inductive effects of H $vs.$ CH_3.

The present analysis focusses on the similarities in the charge distributions deduced from different theoretical methods. The main difficulty is that the "chemical intuition" is, in general, not formulated in a sufficiently quantitative manner. Here, a detailed consideration of the inductive effects of alkyl groups enables us to translate "chemical intuition" into quantitative arguments.

2. INDUCTIVE EFFECTS OF ALKYL GROUPS

Much of the versatile arrow pushing which is used to represent electron displacements in reacting molecules is difficult, if not impossible, to rationalize in a quantitative manner. It remains, however, that in the interpretation of the chemical behavior of organic molecules important arguments are related to changes in electron densities brought about by substitution. Current arguments, based mainly on kinetic and equilibrium evidence, define the following order of electron releasing ability for the alkyl groups[3,4]:

$$tert\text{-}C_4H_9 > iso\text{-}C_3H_7 > C_2H_5 > CH_3. \qquad (2.1)$$

We can now examine to what extent this order of electron-releasing ability is reflected by theoretical population analyses. A justification for this type of approach can be found in the following example. Propane can be viewed as an ethyl group (the better donor) attached to a methyl group. Therefore, provided that the above inductive order is reflected in the electron distributions of ground state molecules, it follows that each methyl group should carry a net negative charge. This is, indeed, the case, as shown by direct experimental evidence[5]. Moreover, both methyl groups being equivalent, the CH_2 group should be electron deficient because the molecule is electroneutral. This is typically a situation which should be reflected by quantum theory, if any meaning at all is to be given to theoretical electron distributions.

A number of qualitative predictions can be derived for a number of alkanes from the inductive order (Eq. 2.1), in a way similar to that employed for predicting that the methyl groups in propane are negatively charged. For example, the net electron loss of a $tert$-butyl group is more important than that of a methyl group under similar circumstances, i.e., if they are attached to the same atom or group of atoms. In neopentane, which can be considered as a $tert$-butyl attached to a methyl group, a net positive charge should therefore be carried by the $tert$-butyl group, and an equal negative charge by the methyl group. Moreover, a net positive charge should result on the quaternary carbon atom, because the four methyl groups are equivalent and the molecule is electroneutral. In isobutane, on the other hand, the net negative charge on the CH_3 groups is expected to be lower than in the case

TABLE 2.1. Polar σ* Constants

Group	σ*	Group	σ*
CH_3	0	neo-C_5H_{11}	−0.151
C_2H_5	−0.100	i-C_3H_7	−0.190
n-C_3H_7	−0.115	sec-C_4H_9	−0.210
n-C_4H_9	−0.124	$(C_2H_5)_2CH$	−0.225
i-C_4H_9	−0.129	tert-C_4H_9	−0.300

From a consideration of all the trends exhibited by the alkyl groups, the slightly modified σ* values for the n-butyl and isobutyl groups appear to be more consistent than the usual ones (−0.130 and −0.125, respectively). In the present ordering, any R—CH_2 group is a better electron donor when R is a better donor, whereas the usual σ* values for n-C_4H_9 and iso-C_4H_9 would indicate an inversion in the ordering for these groups. In any case, the σ* values used here are well within their error limits. This applies also to the σ* value for the neopentyl group which is used here (lit.[4]: −0.165; lit.[3]: −0.14, as determined from the kinetics of sulfation of alcohols). All other σ* values are those of Refs. 3, 4. Finally, the prime importance of the σ*'s does not rest in their exact numerical values (which may, in the future, be subject to minor revisions) but in the fact that the same set of σ*'s is used in all the comparisons with the different theoretical results.

of neopentane, because isobutane is in fact an isopropyl group attached to methyl, and the isopropyl group is a less good electron donor than the tert-butyl group. Finally, while the net charge on the methyl group in propane is still lower (because ethyl is a less efficient electron donor than iso-C_3H_7), in ethane both methyl groups are necessarily electroneutral.

Important quantitative aspects are associated with the inductive order expressed in Eq. 2.1. Indeed, numerous correlations involving experimental (kinetic and equilibrium) data have led to a numerical scaling, in terms of polar σ* substituent constants, of the electron-releasing abilities of the alkyl groups. In Taft's scale[3,4], the electron-releasing ability of the methyl group is assigned the value $\sigma^*(CH_3) = 0$. The ethyl group, which is a better electron donor than methyl, is described by $\sigma^*(C_2H_5) = -0.100$. These two values define the origin and the slope for the evaluation of the inductive effects. In this arbitrary scale, indicated in Table 2.1, the σ* is increasingly negative as the group it describes is a better electron donor.

Tentatively assuming that the Taft polar constants do, indeed, "measure" electron-releasing abilities, we shall now examine whether and how these electronic effects are reflected in population analyses calculated from conventional theoretical methods. This attempt is expected to succeed at least in the case of propane because, as mentioned above, direct experimental evidence indicates an actual electron enrichment at the methyl groups. Selected results, both semi-empirical and ab initio, are presented in Table 2.2. A vivid illustration of the results given in Table 2.2 is presented in Figure 2.1. The net charges of the methyl groups, $q(CH_3)$, in selected alkanes R—CH_3 are compared with the electron-releasing abilities of the R groups, which

TABLE 2.2. Net Charges of Methyl Groups in R—CH$_3$ Compounds

R	$q(CH_3)$, me		
	EHMO	PCILO	STO–3G
CH$_3$	0	0	0
C$_2$H$_5$	−13	−7.4	−5.15
i-C$_3$H$_7$	−25	−15.0	−9.91
tert-C$_4$H$_9$	−35	−20.0	−15.73

Results extracted from Refs. 6–8.

are expressed by the σ^* values of Table 2.1. Two points are noteworthy: (*i*) the correlation of the theoretical methyl net charges with the "usual" inductive effects (as they are known from a large body of experimental information) and, more importantly, (*ii*) that the same type of correlation is obtained from different sets of theoretical results meaning, hence, that they are similar as far as this particular aspect is concerned.

The arguments which were developed so far with reference to alkanes, simply written as R—CH$_3$, apply equally well to this set of molecules, now written as R—H. For example, since the ethyl group is a better electron donor than methyl, any H atom in ethane is expected to be electron richer than a methane H atom. Moreover, the tertiary H atom in isobutane is expected to carry more negative charge than any ethane H atom, because the *tert*-butyl group is a better electron donor than the ethyl group. Finally, the net charge of a secondary H atom in propane is intermediate between that of the tertiary hydrogen in isobutane and that of an ethane H atom.

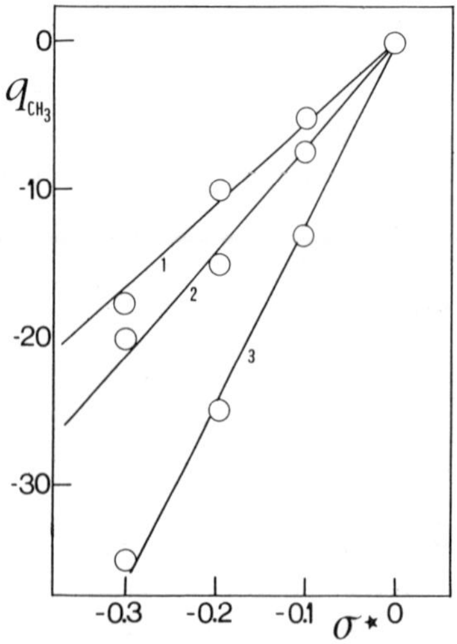

FIGURE 2.1. Net charge of methyl groups, in 10^{-3} electron units, *vs.* Taft's polar constants, σ^*. The charge results are from STO–3G (1), PCILO (2), and EHMO (3) calculations.

TABLE 2.3. Hydrogen Net Charges in R—H Compounds

	q_H, me		
R	EHMO	PCILO	STO–3G
CH_3	126	10.8	12.23
C_2H_5	109	3.5	6.99
i-C_3H_7	95	−3.9	2.20
$tert$-C_4H_9	82	−10.5	−3.53

Results extracted from Refs. 6–8.

These expectations are, indeed, well-reflected in theoretical population analyses, as indicated by the results displayed in Table 2.3 and Figure 2.2.

The information contained in Figures 2.1 and 2.2 represents the clue from which a general approach can be developed for the charge distributions in saturated hydrocarbons. At this stage, it is also interesting to note that correlations of similar type, involving polar σ^* constants, are valid for alkyl groups attached to ethylenic[9,10] and phenyl groups[9]. It may be concluded that population analyses calculated for a variety of compounds by means of different theoretical methods, both semi-empirical and *ab initio*, reflect the electron-releasing ability of the alkyl groups in Taft's order, suggesting that the usual interpretation of inductive effects in terms of electron release is adequate. The important point is that all theoretical methods are equivalent regarding this particular aspect of inductive effects, despite the fact that the charges of the individual atoms are not the same when they are derived by different methods[6]. The idea suggested by Figs. 2.1 and 2.2 is simple. Since

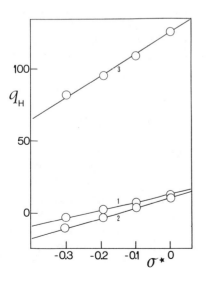

FIGURE 2.2. Hydrogen net charges, in 10^{-3} electron units, *vs.* Taft's polar constants, σ^*. The charges are deduced from STO–3G (1), PCILO (2), and EHMO calculations.

quantum mechanical charges derived from different theoretical models reflect the inductive effects of alkyl groups, we can use the latter as a starting point for a back-calculation of atomic charges which should, in principle, reproduce the theoretical sets of results. This type of calculation is made as follows[6].

3. BACK-CALCULATION OF ATOMIC CHARGES FROM THE INDUCTIVE EFFECTS

Correlations between atomic charges and inductive effects are significant in that they are quite general, although the theoretical methods used in this type of comparison differ from one another in the models underlying them and inductive effects are in no case specifically built in. While one may or may not be interested in the inductive effects themselves, the fact remains that the charges derived from different methods are similar in one respect, i.e., in that they follow certain rules (Figs. 2.1 and 2.2) expressed in terms of a fixed set of numbers (the σ^* parameters). This is, in fact, the only point that matters for the following calculations.

The linear relationship, Fig. 2.1, between $q(CH_3)$ and σ_R^* is expressed in Eq. 2.2 for alkanes written as $R\!-\!CH_3$.

$$q(CH_3) = a\sigma_R^*. \tag{2.2}$$

The slope, a, represents the sensitivity to substitution and its numerical value differs from one method to another. No constant term appears in Eq. 2.2 because $q(CH_3) = 0$ for $\sigma_R^* = \sigma^*(CH_3) = 0$, which is the case for ethane. Similarly, it follows from Fig. 2.2 (for alkanes written as $R\!-\!H$) that

$$q_H = a\sigma_R^* + b. \tag{2.3}$$

For $R = CH_3$ (i.e., $\sigma_R^* = 0$), Eq. 2.3 gives $b =$ hydrogen net charge in CH_4. Within the precision of this type of analysis, semi-empirical charge results show that the slope, a, is the same in both Eqs. 2.2 and 2.3, and accurate *ab initio* calculations involving full optimization of geometry and of all individual ζ exponents lead precisely to the same result. This point is best illustrated by the numerical comparisons presented further below (Sect. 4). Finally, Eq. 2.4 expresses the fact that the molecules are electroneutral.

$$\sum q_k = 0. \tag{2.4}$$

Equations 2.2–2.4 enable the back-calculation of charge distributions from the knowledge of the inductive effects, i.e., more precisely, from the polar substituent constants σ^*. The general outline of such calculations can be illustrated as follows, taking propane as an example. In this molecule there are four unknown charges, namely, the primary and secondary carbon atoms and two different hydrogen atoms (the weighted average of the primary H atoms is considered in this case). Because of the charge normalization constraint (Eq. 2.4), three equations are required for solving the problem of the three remaining unknown charges. One of these equations is given by Eq. 2.2 considering propane as C_2H_5—CH_3, the other two are given by Eq. 2.3, considering propane as n-C_3H_7—H (for the calculation of the primary H atoms) or as iso-C_3H_7—H (for the secondary H atoms). Thus, the problem can be solved. Of course, this type of approach is by no means a general way of obtaining charges but the fact that it can be applied to an adequate collection of molecules is sufficient for our intended purpose.

There is one question which must be dealt with at this point, namely, the choice of the appropriate unit for net atomic charges. A suitable unit is defined by selecting the carbon atom of ethane as reference and by taking its net charge at 1 unit. In this system of "relative units", the hydrogen net charge in ethane is $-1/3 = b - 0.1a$ (from charge normalization, $viz.$, Eq. 2.3) where, of course, a and b are also expressed in "relative units". In order to obtain atomic charges (as well as a and b) in terms of, say, 10^{-3} electron units, it suffices to multiply the results expressed in "relative units" by the numerical value of

$$q_C^\circ = \text{carbon net charge in ethane} \qquad (2.5)$$

expressed in charge units (e.g., in 10^{-3} electron units). Moreover, we choose to express the slope a as follows

$$a = -\frac{10}{3n} \quad \text{("relative units")} \qquad (2.6)$$

in terms of the new variable n because this change of variable turns out to be a handy one at the level of the final results. Of course, if a is to be expressed in charge units, we write $a = -(10/3n)q_C^\circ$ (charge units). The physical meaning of n can be inferred from Eq. 2.6 where a measures, in a way, the sensitivity of the variations in charge due to substituent effects (Eqs. 2.2 and 2.3). Small n values indicate strong substituent effects. At the other limit, if inductive effects were nonexistent, the charge distributions would be those corresponding to $n = \infty$ (i.e., $a = 0$); in that event, however, the hydrogen net charges would be the same in ethane and methane (Eq. 2.3), meaning that in the absence of inductive effects all net charges should be 0 in order to satisfy the constraint of molecular electroneutrality. Of course, none of the various theoretical methods leads to this extreme result: they differ, however, from one another by the value of n.

At this stage it appears interesting to consider the sign of a. From Eq. 2.3 it is readily deduced that a must be positive in order to satisfy the usual Taft order of electron-releasing abilities $tert$-$C_4H_9 > \cdots > CH_3$ because only then will a hydrogen atom be electron richer when attached to a $tert$-butyl group than in methane. Similarly, only then (Eq. 2.2) will the methyl group in propane carry a net negative charge, as indicated by experiment[5]. Moreover, writing now a in charge units $(-10q_C^\circ/3n)$, it is also deduced that q_C° and n are opposite in sign. Since by definition (Eq. 2.5) q_C° is the carbon net charge in ethane, it follows that the inductive order requires that n is positive if the polarity is $C^- — H^+$ and negative for $C^+ — H^-$. Most of the theoretical results presented further below correspond to $n > 0$, with $C^- — H^+$. Optimized STO–3G Mulliken charges[7], for example, correspond to $q_C^\circ = -20.96$ me and $n = 1.3325$. The INDO results, as well as SCF–Xα–SW charges[12], are exceptional because they correspond to negative n values, -2 and -4.4293, respectively, the ethane-C net charge being positive in both cases.

The carbon and hydrogen net charges in a series of alkanes are now readily deduced in the following manner. In relative units, the ethane-H net charge is $-1/3 = b - 0.1a$ (Eq. 2.3); hence, using Eq. 2.6, we obtain

$$b = -\frac{n+1}{3n} \quad \text{(rel. un.).} \tag{2.7}$$

It follows that the carbon net charge in methane $(= -4b$, from charge normalization) is simply $4(n + 1)/3n$ (rel. un.) or $[4(n + 1)/3n]q_C^\circ$ charge units. Using this expression for b and Eq. 2.6 together with Eq. 2.3, we find a general formula for the hydrogen net charges:

$$q_H = -\frac{10\sigma^* + n + 1}{3n} \quad \text{(rel. un.).} \tag{2.8}$$

Finally, using the charge normalization constraint (Eq. 2.4) and writing now Eq. 2.2 in its alternate form

$$q(CH_3) = -\left(\frac{10}{3n}\right)\sigma^* \quad \text{(rel. un.)} \tag{2.9}$$

we obtain a set of formulas describing atomic charges in terms of n by employing the appropriate σ^*'s given in Table 2.1. These formulas[6] are indicated in Table 2.4.

The results expressed by these charge formulas are quite general in that they reflect, in principle, net charges given by any theoretical method. It is, indeed, sufficient to know what particular n value corresponds to a given theoretical approach [e.g., from the theoretical ratio $q_C(\text{methane})/q_C(\text{ethane})$] so that we can generate the corresponding charges for all the molecules of the set, in relative units. A multiplication of the results obtained in this fashion by the appropriate carbon net charge in ethane yields atomic populations which are now expressed in terms of charge units. The important point is that Table 2.4 represents, in fact, a summary of atomic charges given by any theoretical method reproducing the inductive effects, thus permitting a general discussion of atomic charges without referring in particular to any one of the various theoretical approaches. That this study can be made only for a limited number of molecules is no restriction in this case.

Of course, one may regard the last statements as being perhaps somewhat strong. Now we must back them up.

TABLE 2.4. Net Charges in Relative Units

Molecule	Atom	Net Charge
Methane	C	$4(n + 1)/3n$
Ethane	C	1.000
Propane	C_{prim}	$(3n + 0.55)/3n$
	C_{sec}	$(2n - 3.8)/3n$
	H_{prim}	$(0.15 - n)/3n$
	H_{sec}	$(0.9 - n)/3n$
Butane	C_{prim}	$(3n + 0.43)/3n$
	C_{sec}	$(2n - 3.35)/3n$
	H_{prim}	$(0.24 - n)/3n$
	H_{sec}	$(1.1 - n)/3n$
Pentane	C_{centr}	$(2n - 2.8)/3n$
	H_{centr}	$(1.25 - n)/3n$
Isobutane	C_{prim}	$(3n + 1.03)/3n$
	C_{tert}	$(n - 7.7)/3n$
	H_{prim}	$(0.29 - n)/3n$
	H_{tert}	$(2 - n)/3n$
Neopentane	C_{prim}	$(n + 0.49)/n$
	C_{quat}	$-4/n$
	H	$(0.51 - n)/3n$

The net charge of the central CH_2 group in *n*-pentane was calculated assuming additivity of the inductive (σ^*) effects, i.e., $q(C_2H_5) = a[\sigma^*(n\text{-propyl}) - \sigma^*(\text{ethyl})]$, and $q(CH_2) = -2q(C_2H_5)$.

4. COMPARISON WITH THEORETICAL POPULATION ANALYSES

The following comparisons between quantum mechanical charge distributions and their "inductive" counterparts of Table 2.4 involve theoretical charges deduced from conventional methods which are well described in the literature. The appropriate n and q_C° values were determined for each theoretical method in order to achieve the best possible agreement with the "inductive" charges of Table 2.4.

The comparison involving charges calculated by Del Re's method[13] proves satisfactory with $n = 34$ and $q_C^\circ = -117$ me. A similar agreement is also found for Mulliken charges calculated by Hoffmann[1,6], with $n = 9.5$ and $q_C^\circ = -356$ me. With CNDO/2 results[8,14] the situation is, however, a little different. Indeed, while the CNDO/2 charges are clearly in qualitative agreement with the "inductive" results in that they reproduce correctly all the predicted major trends, the actual numerical agreement appears to be somewhat erratic. An "agreement" of similar sort between "inductive" and theoretical charges is also observed in the results obtained by Salahub and

Sándorfy[11], using a modified CNDO approach; the latter is a variant of that of Del Bene and Jaffé[15], which was modified to interpret correctly σ-electronic spectra. A calculation of CNDO type, including higher (Rydberg) atomic orbitals in the basis, was also presented by Salahub and Sándorfy[11]. In this RCNDO method, H $2s$, H $2p$, C $3s$, and C $3p$ Slater atomic orbitals were added to the basis. In these calculations the energy of the ground state was first minimized in the usual way, then configuration interaction was applied including the lowest 30 singly excited configurations for both singlets and triplets. As indicated in Table 2.5, the agreement of the RCNDO charges with the "inductive" ones is satisfactory and is superior to what can be obtained from the original CNDO/2 method.

The INDO charge distributions also calculated by Salahub and Sándorfy[11] are compared in Table 2.5 with the charges of Table 2.4, with $n = -2$ and $q_C^\circ = 71$ me. Contrasting with all the other semi-empirical results discussed here, INDO calculations predict a C^+—H^- polarity. Except for neopentane, where the INDO charge for the quaternary C atom appears to be less positive than that of the tertiary C atom in isobutane (which is contrary to the results given by all other methods), it appears that INDO charges reflect correctly the major trends predicted by the "inductive" formulas in spite of this change in CH polarity. A basically different theoretical model is that proposed by Hoffmann[1]. A variant of his "Extended Hückel" (EHMO) method was investigated by Sichel and Whitehead[16]; their results are in satisfactory agreement with the inductive formulas, with $n = 9.14$ and $q_C^\circ = -338$ me, as indicated in Table 2.5. Similarly, the PCILO (Perturbative Configuration Interaction of Localized Orbitals[17]) charge distributions compare favorably with their "inductive counterparts[8], with $n = 0.525$ and $q_C^\circ = -11$ me (Table 2.5). We may conclude that charge analyses derived by means of semi-empirical methods are in satisfactory general agreement with their counterparts predicted from the inductive effects of alkyl groups.

On the other hand, the SCF-type results obtained by Lipscomb et al.[18] also appear to follow the general trends predicted by the "inductive approach" (Table 2.6). Finally, the SCF–Xα–SW method of Slater and Johnson[19], which does not involve a Mulliken population analysis, also gives satisfactory results with $n = -4.4293$ and $q_C^\circ = 612$ me[12].

Let us now examine sets of charges calculated by *ab initio* methods of different degrees of sophistication. "Standard" STO–3G calculations[7] following Pople's recipe[20], assuming standard geometries and a fixed set of ζ exponents, yield the crudest possible set of *ab initio* results. The Mulliken charges given in Table 2.7 are nevertheless in satisfactory agreement with those derived from the "inductive" formulas of Table 2.4, with $n = 1$ and $q_C^\circ = -26$ me; the largest discrepancy is observed for the quaternary C atom in neopentane but, as indicated by more sophisticated *ab initio* calculations, this departure should not be considered as indicative of a disagreement for which the "inductive" approach can be made responsible.

In STO–NG calculations, total molecular energies converge rapidly

TABLE 2.5. A Comparison of Semi-empirical Theoretical Results with the Charges Deduced from Table 2.4 (10^{-3} eu)

Molecule	Atom	RCNDO	Inductive $n = 1.6$	INDO	Inductive $n = -2$	PCILO	Inductive $n = 0.525$	EHMO	Inductive $n = 9.14$
Methane	C	−84	−82	+43	+47	−43.3	−42.6	−503	−500
Ethane	C	−34	−38	+77	+71	−10.4	−11.0	−326	−338
Propane	C_{prim}	−44	−42	+67	+65	−14.6	−14.8	−343	−345
	C_{sec}	+12	+5	+94	+92	+20.2	+20.6	−164	−178
	H_{prim}			−25	−25	+2.0	+2.6	+110	+111
	H_{sec}			−37	−34	−3.9	−3.3	+95	+102
Butane	C_{prim}	−47	−42	+65	+66			−342	−343
	C_{sec}	+2	+1	+85	+87			−181	−184
	H_{prim}			−26	−26			+110	+110
	H_{sec}			−36	−37			+96	+99
Pentane	C_{centr}			+76	+80				
	H_{centr}			−36	−38				
Isobutane	C_{prim}	−52	−47	+62	+59	−19.6	−19.7	−358	−351
	C_{tert}	+53	+48	+103	+115	+51.3	+52.2	−8	−18
	H_{prim}			−27	−27	+1.1	+1.9	+111	+109
	H_{tert}			−48	−47	−10.5	−10.3	+82	+88
Neopentane	C_{prim}	−57	−50	+58	+54	−19.9	−18.4	−373	−356
	C_{quat}	+92	+95	+97	+142	+82.0	+83.8	+140	+148
	H			−27	−30	+0.2	−0.9	+113	+106

TABLE 2.6. A Comparison of SCF-Type and SCF–Xα–SW Charges with their "Inductive" Counterparts (electron units)

Molecule	Atom	SCF	Inductive $n = 1.6964$	SCF–Xα–SW	Inductive $n = -4.4293$
Methane	C	−0.49	−0.49	0.63	0.63
Ethane	C	−0.19	−0.23	0.61	0.61
Propane	C_{prim}	−0.25	−0.25	0.60	0.59
	C_{sec}	−0.05	0.02	0.59	0.58
	H_{prim}			−0.21	−0.21
	H_{sec}			−0.25	−0.25
Butane	C_{prim}	−0.26	−0.25		
	C_{sec}	0.01	0.00		
Pentane	C_{centr}	−0.03	−0.03		
Isobutane	C_{prim}	−0.28	−0.28	0.56	0.56
	C_{tert}	0.30	0.27	0.56	0.56
	H_{prim}			−0.22	−0.22
	H_{tert}			−0.29	−0.30
Neopentane	C_{prim}	−0.29	−0.29	0.54	0.54
	C_{quat}	0.54	0.54	0.55	0.55

towards the full STO result with increasing size of Gaussian expansions[20–22]. Improvement of inner-shell description, e.g., at the 6-31G level[22], further improves molecular energy calculations. The point is that any such improvement in total molecular energy does not necessarily mean that all calculated properties are drastically different or, simply, better than what can be obtained from the simple STO–NG calculations. A substantial lowering of the energies with respect to STO–3G energies is observed in GTO $(6s3p|3s)$ calculations[7]. A comparison of these Mulliken GTO charges with their "inductive" counterparts is presented in Table 2.7. The agreement is satisfactory, with $n = 14.11$ and $q_C^\circ = -232.5$ me. Larger charge separations are obtained from a $(7s3p|3s)$ expansion[23]; the results correspond to a very high n value (42.316) and are also satisfactorily reproduced by the "inductive" method. As is the case with these $(7s3p|3s)$ calculations, the charge separations derived by Hoyland[24] using a bond order (BO) method are much larger than those deduced from a minimal basis set. Again, the extended basis set results correspond to a higher n value ($n = 25$) than that corresponding to minimal basis set calculations. With these BO charges, the q_C° value is also significantly larger, $q_C^\circ = -504$ me, than that deduced from simple STO–3G calculations. The agreement with the "inductive" charges appears to be satisfactory.

With the *ab initio* methods used for deducing the charges given in Table 2.7, improvement of the theoretical results is achieved, at least from the point of view of total molecular energies, by expansion of the Gaussian

TABLE 2.7. A Comparison of *Ab Initio* Results with Charges Deduced from Table 2.4

Molecule	Atom	STO–3G[a] "standard"	Inductive[a] n = 1	GTO(6s3p/3s)[a]	Inductive[a] n = 14.11	7s3p/3s[b]	Inductive[b] n = 42.316	BO[a]	Inductive[a] n = 25
Methane	C	−72	−69	−334.8	−331.5	−0.79	−0.78	−699	−699
Ethane	C	−26	−26	−234.0	−232.5	−0.57	−0.57	−504	−504
Propane	C_{prim}	−31	−31	−236.5	−235.6	−0.58	−0.58	−508	−508
	C_{sec}	+19	+16	−132.9	−134.6	−0.38	−0.37	−303	−310
	H_{prim} [c]	+7	+7.4	+75.9	+76.7			+167	+167
	H_{sec}	+0.3	+0.9	+75.3	+72.6			+160	+162
Butane	C_{prim}					−0.56	−0.57	−504	−507
	C_{sec}					−0.37	−0.37	−309	−313
	H_{prim}							+165	+166
	H_{sec}							+159	+160
Isobutane	C_{prim}	−35	−35	−238.3	−238.0	−0.55	−0.58		
	C_{tert}	+57	+58	−27.6	−36.1	−0.18	−0.16		
	H_{prim}	+6	+6.2	+74.5	+75.9				
	H_{tert}	−6.2	−8.7	+71.8	+66.5				
Neopentane	C_{prim}	−39	−39	−237.3	−240.4				
	C_{quat}	+93	+104	+60.1	+64.5				

[a] 10^{-3} electron units. [b] Electron units. [c] Average of nonequivalent H atoms.

TABLE 2.8. A Comparison of Optimized STO–3G Results with the "Inductive" Charges Deduced from Table 2.4 (me)

Molecule	Atom	STO–3G "optimized"	Inductive n = 1.3325
Methane	C	−48.92	−48.92
Ethane	C	−20.96	−20.96
Propane	C_{prim}	−23.81	−23.84
	C_{sec}	+5.94	+5.95
	H_{prim}[†]	+6.23	+6.20
	H_{sec}	+2.20	+2.27
Isobutane	C_{prim}	−26.39	−26.36
	C_{tert}	+33.36	+33.39
	H_{prim}[†]	+5.50	+5.47
	H_{tert}	−3.53	−3.50
Neopentane	C_{prim}	−28.66	−28.67
	C_{quat}	+62.92	+62.92

[†] Weighted average of nonequivalent primary H atoms.

descriptions. The agreement between "inductive" and quantum mechanical charges, however, appears to be only marginally affected by the use of increasing Gaussian descriptions. On the other hand, there is an important way of improving charge calculations, even within the framework of the simplest possible *ab initio* scheme, that of the STO–3G method. Indeed, the study of the ζ exponents when treated as variational parameters indicates that the optimized values (for each type of atom) are largely independent of the size of the Gaussian expansion but vary considerably from one molecule to another. Now it is known that charge distributions are very sensitive to the molecular environment and, hence, to ζ optimization. Consequently, any attempt to calculate net charges which may be regarded "accurate" at the level of the method which is used to derive them implies a very careful optimization of the exponents and, to a lesser extent, of the molecular geometry. Not surprisingly, more accurate minimal basis set calculations carried out along these lines reveal a considerable improvement over the "standard" STO–3G results described in Table 2.7. In this case we use optimized molecular geometries and all exponents are optimized individually for each molecule, including those of the carbon K shells. Of course, different C and H atoms in the same molecule are optimized separately, and each energy minimization is carried out until stable atomic charges are obtained, within ~0.01 me. As indicated in Table 2.8, the Mulliken net charges obtained[7] from this "best" calculation using a minimal basis set are virtually identical to their "inductive" counterparts, with $n = 1.3325$ and $q_C^\circ = -20.96$ me.

5. CONCLUSIONS

The usual interpretation of the inductive effects of alkyl groups in terms of electron release appears to be adequate as its description is, in general, satisfactorily reproduced by quantum mechanical charge distributions. This result is significant because inductive effects, in their usual interpretation, are not specifically built in as a part of quantum theory, and is important because of the experimental evidence indicating that the methyl groups in propane carry a net negative charge.

In alkanes, the net atomic charges derived from the inductive Taft-like equations $q_H = a\sigma_R^* + b$ and $q(CH_3) = a\sigma_R^*$ (for R—H, viz., R—CH$_3$ compounds) can be expressed in terms of one parameter, $n = -10/3a$, in a scale of relative units defined by $q_C(C_2H_6) = 1$ (rel. un.). These "inductive" charges are given in Table 2.4. By multiplying them by q_C° (i.e., the carbon net charge of ethane, expressed in charge units), electron populations of the individual atoms are obtained in charge units. The quantities q_C° and n are opposite in sign. Positive n values correspond to a C$^-$—H$^+$ polarity, whereas $n < 0$ corresponds to C$^+$—H$^-$. A comparison of these "inductive" charge distributions with 15 different sets of theoretical results, derived from semi-empirical or partially or nonoptimized *ab initio* quantum mechanical methods, reveals an overall agreement which is satisfactory. With fully optimized *ab initio* (STO–3G) results (in principle, the best calculation using a minimal basis set), the agreement is virtually perfect (Table 2.8).

TABLE 2.9. Summary of the "Inductive" Parameters, n and q_C°, which Generate the Charge Results Given by Theoretical Calculations[6]

Method	n	q_C° (me)
INDO	−2	+71
CNDO/2	0.35	−7.6
PCILO	0.525	−11
RCNDO	1.6	−38
PPP-type	9.4	−160
EHMO	9.14	−338
EHMO	9.5	−356
DelRe	34	−117
SCF–Xα–SW	−4.4293	+612
STO–3G "standard"	1.0	−26
STO–3G "optimized"	1.3325	−20.96
SCF-type	1.7	−230
BO	25	−504
$(6s3p|3s)$	14.11	−232.5
$(7s3p|3s)$	42.3	−573

These results illustrate the flexibility contained in the inductive approach. The equations of Table 2.4 represent any possible scheme of charge distributions, ranging from C^+—H^- to C^-—H^+ situations, depending upon the choice of n, whereas any individual theoretical method represents only one case, corresponding to a specified value of n. The set of "inductive" charge distributions (Table 2.4) represents, hence, a summary of sets of theoretical net charges obtained from the various theoretical approaches, each of which can be generated by employing the appropriate n and q_C° values indicated in Table 2.9.

It is also clear, however, that the "inductive theory" is by no means a general way for obtaining charge distributions, as it is perforce limited to molecules which can be constructed from fragments whose polar σ^* values are known and to comparisons involving the replacement of one alkyl group by another. Fortunately, this restriction represents no difficulty in the development of a more general approach based on the inductive formulas (considered as a summary of theoretical results) and on Eqs. 1.3–1.5. Finally, two important points remain to be clarified, namely, the physically "true" value of n, a problem which is related to the mode of partitioning overlap populations in Mulliken-type analyses, and that of q_C°.

REFERENCES

1. R. Hoffmann, *J. Chem. Phys.*, **39**, 1397 (1963).
2. W.J. Hehre and J.A. Pople, *J. Am. Chem. Soc.*, **92**, 2191 (1970).
3. R.W. Taft, *J. Am. Chem. Soc.*, **75**, 4231 (1953).
4. R.W. Taft, in "Steric Effects in Organic Chemistry", M.S. Newman, Ed., Wiley, New York, NY, 1956.
5. V.W. Laurie and J.S. Muenter, *J. Am. Chem. Soc.*, **88**, 2883 (1966).
6. S. Fliszár, G. Kean, and R. Macaulay, *J. Am. Chem. Soc.*, **96**, 4354 (1974).
7. G. Kean and S. Fliszár, *Can. J. Chem.*, **52**, 2772 (1974).
8. S. Fliszár and J. Sygusch, *Can J. Chem.*, **51**, 991 (1973).
9. J. Grignon and S. Fliszár, *Can. J. Chem.*, **52**, 2766 (1974).
10. H. Henry and S. Fliszár, *J. Am. Chem. Soc.*, **100**, 3312 (1978).
11. D.R. Salahub and C. Sándorfy, *Theor. Chim. Acta*, **20**, 227 (1971).
12. A.E. Foti, V.H. Smith, and S. Fliszár, *J. Mol. Struct.*, **68**, 227 (1980).
13. G. Del Re, *J. Chem. Soc.*, 4031 (1958); S. Fliszár, *J. Am. Chem. Soc.*, **94**, 7386 (1972).
14. J.A. Pople and G.A. Segal, *J. Chem. Phys.*, **44**, 3289 (1966).
15. J. Del Bene and H.H. Jaffé, *J. Chem. Phys.*, **48**, 1807 (1968).
16. J.M. Sichel and M.A. Whitehead, *Theor. Chim. Acta*, **5**, 35 (1966).
17. S. Diner, J.P. Malrieu, and P. Claverie, *Theor. Chim. Acta*, **13**, 1 (1969); J.P. Malrieu, P. Claverie, and S. Diner, *Theor. Chim. Acta*, **13**, 18 (1969); S. Diner, J.P. Malrieu, F. Jordan, and M. Gilbert, *Theor. Chim. Acta*, **13**, 101 (1969); J. Langlet, B. Pullman, and M. Berthod, *J. Mol. Struct.*, **6**, 139 (1970); B. Maigret, B. Pullman, and J. Caillet, *Biochim. Biophys. Res. Commun.*, **40**, 808 (1970); J. Langlet, B. Pullman, and H. Berthod, *J. Chim. Phys.*, **67**, 480 (1970).
18. M.D. Newton, F.P. Boer, and W.N. Lipscomb, *J. Am. Chem. Soc.*, **88**, 2367 (1966).

19. J.C. Slater, *Adv. Quantum Chem.*, **6**, 1 (1973); K.H. Johnson, *Adv. Quantum Chem.*, **7**, 143 (1973); K.H. Johnson, *Ann. Rev. Phys. Chem.*, **26**, 39 (1975); J.W.D. Connolly, in "Modern Theoretical Chemistry", G.A. Segal, Ed., Vol. IV, Plenum, New York, NY, 1977.
20. W.J. Hehre, R.F. Stewart, and J.A. Pople, *J. Chem. Phys.*, **51**, 2657 (1969).
21. L. Radom and J.A. Pople, in *MTP International Review of Science, Physical Chemistry*, Series One, Vol. 1, W. Byers Brown, Ed., Butterworths and Co. Ltd., London, 1972, p. 71.
22. W.J. Hehre, R. Ditchfield, and J.A. Pople, *J. Chem. Phys.*, **56**, 2257 (1972).
23. J.M. André, P. Degand, and G. Leroy, *Bull. Soc. Chim. Belg.*, **80**, 585 (1971).
24. J.R. Hoyland, *J. Chem. Phys.*, **50**, 473 (1969).

CHAPTER 3

A Modified Population Analysis

1. CRITERION FOR SELECTING A THEORETICAL METHOD FOR THE STUDY OF MOLECULAR PROPERTIES INVOLVING CHARGES

The comparison of quantum mechanical charge distributions with their "inductive" counterparts, $f(n, q_C^\circ)$, reveals a strong common point relating theoretical charges derived in a variety of different manners. Indeed, although theoretically deduced results cover a wide spectrum of numeral values (ranging from C^+—H^- to C^-—H^+ situations) supposedly describing net charges of selected atoms, each individual theoretical method yields a set of results reflecting the customary inductive effects of alkyl groups in a quite consistent fashion. In short, the set of "inductive" charges $f(n, q_C^\circ)$ represents a summary of sets of theoretical results. Of course, the n and q_C° values to be used for reproducing theoretical charges differ from case to case (Table 2.9). This situation now raises the following questions.

Suppose we were to study a correlation between an observable property (e.g., carbon–13 NMR shifts or a molecular energy) and calculated atomic charges. The first question would then concern the choice of the theoretical method to be used for this comparison. Present results indicate that *any choice of any particular theoretical method in fact reduces to choosing a particular n value* (i.e., that corresponding to the selected method). While all sorts of arguments may be invoked for justifying the preference for one method over the others, it appears difficult to justify *a priori* why any particular n value should be preferred for describing the charges to which the property under observation should be correlated. This leads to the obvious question about the physically "true" value of n, which shall now be discussed keeping in mind possible comparisons with experimental data.

FIGURE 3.1. A comparison of carbon net charges, in relative units, calculated (from Table 2.4) for $n = 5$ with their counterparts calculated for $n = -4.4$. (Results extracted from Ref. 1).

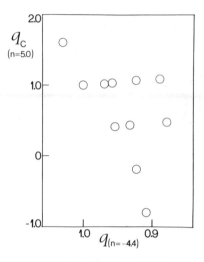

To begin with, let us examine a hypothetical property which we suppose to be linearly related to q_H. From Eq. 2.8 (expressed in charge units), i.e.,

$$q_H = -\frac{10\sigma^* + n + 1}{3n} q_C^\circ$$

it follows that the hydrogen net charges calculated for one given value of n are linearly related to their counterparts corresponding to another n value. Consequently, such a hypothetical property is linearly related to hydrogen net charges given by any theoretical method, independently of its corresponding n value, provided only that the theoretical results are in satisfactory agreement with the set of "inductive" charges. In short, any good theoretical method can be used for the study of a property-charge relationship involving hydrogen atoms, as the corresponding n value is by no means of crucial importance.

The situation is, however, different for the charges on carbon atoms, i.e., in any study of molecular properties involving them. This statement is best demonstrated with the aid of Fig. 3.1, which illustrates the fact that carbon net charges calculated for one given value of n (e.g., $n = 5$) are not linearly related to their counterparts deduced for any other n values (e.g., $n = -4.4$).

Consider now a hypothetical property which is supposedly linearly related to carbon net charges and examine the comparison (Fig. 3.1) between C net charges calculated for two different values of n. Obviously, such a property cannot be related linearly to *both* sets of carbon charges (e.g., to those corresponding to $n = -4.4$ *and* to those for $n = 5$) as the scaling and the relative ordering of the individual C charges depend strongly on n. Consequently, *if a property is expected to correlate with carbon net charges, it is imperative that the latter correspond to the "proper" n value: the use of any theoretical set of charge distributions corresponding to an n value other than the appropriate one is bound to fail to show the expected correlation*[1].

The final step involves understanding the physical meaning of n and evaluating its "true" value. Anticipating the forthcoming results, we note that n can be related very simply to the mode of partitioning overlap populations. In this sense, the flexibility contained in the set of "inductive" equations comes from the fact that they are not confined to any particular artificial mode of partitioning overlap populations.

2. A MODIFIED POPULATION ANALYSIS

Mulliken's population analysis scheme for calculating electron populations $(N_k, N_l, \ldots)$ associated with atoms $k, l, \ldots$ implies halving of all overlap population terms (Eq. 1.1). Numerous modifications of this scheme have been proposed in an attempt to find a more realistic basis for distributing the charges: the major difficulty rests in the proper choice of the criterion guiding the allocation of the overlap populations. While the usual half-and-half assignment is certainly easy to justify in situations involving homonuclear overlap partners, this may be a less good approximation in cases involving dissimilar atoms, namely, for C—H overlap populations and even more so for CO, CF, NH, etc. Notwithstanding this reservation, it appears most convenient (i) to use Mulliken's original scheme for calculating charge distributions (such as those derived from the optimized STO–3G method), thus providing sets of results in a well-defined standard of reference; then (ii) to evaluate modifications in terms of departures from this standard of reference, using Eq. 1.5.

Because of the sharply decreasing values of the overlap integrals S with distance, the principal contributions by far to the sum $\Sigma_{l \neq k} p_{kl}$ (Eqs. 1.4 and 1.5) are those due to the overlap terms involving the atoms l directly bonded to k. Moreover, the overlap populations involving specified kl atom pairs under comparable situations being practically the same in series of closely related compounds, one can use Eq. 1.5 with the assumption of a constant p_{kl} term for each type of kl bond formed by atom k, at least in σ systems. Charge normalization is, of course, ensured by the condition $p_{lk} = -p_{kl}$. In this approximation, we write $\Sigma p_{CH} = N_{CH} p$ for the saturated hydrocarbons under study, where p is the departure from the usual halving of the CH overlap population for one CH bond and N_{CH} is the number of hydrogen atoms bonded to a given carbon atom. Consequently, we obtain in this manner from Eq. 1.5 that[2]

$$q_H = q_H^{Mull} - p \tag{3.1}$$

$$q_C = q_C^{Mull} + N_{CH} p. \tag{3.2}$$

We can now examine in what manner the modified charges q_H and q_C of Eqs. 3.1 and 3.2 compare with their original (Mulliken) counterparts within the scheme of "inductive" charges. Applying Eq. 2.8 in charge units, we write

$$q_H = -\frac{10\sigma^* + 1 + n}{3n}q_C^\circ$$

$$q_H^{Mull} = -\frac{10\sigma^* + 1 + n^{Mull}}{3n^{Mull}}q_C^{\circ Mull}$$

where n^{Mull} and $q_C^{\circ Mull}$ are the appropriate parameters corresponding to the nonmodified hydrogen net charges q_H^{Mull}. The slopes of the hydrogen net charges $vs.$ the corresponding σ^* values are $-10q_C^\circ/3n$ for the modified charges and $-10q_C^{\circ Mull}/3n^{Mull}$ for the original Mulliken charges. It is clear, however, that the transformation expressed in Eq. 3.1 leaves the slope of the hydrogen net charges $vs.$ σ^* unaffected, thus indicating that

$$\frac{q_C^\circ}{n} = \frac{q_C^{\circ Mull}}{n^{Mull}}. \tag{3.3}$$

The quantity p can now be deduced from Eq. 3.1, inserting the above expressions for q_H and q_H^{Mull}, and using also Eq. 3.3. The result

$$p = \frac{q_C^\circ - q_C^{\circ Mull}}{3} \tag{3.4}$$

can also be deduced simply by observing that q_C° and $q_C^{\circ Mull}$ are, respectively, the modified and the original carbon net charges in ethane and that their difference represents the correction for three CH overlap populations (Eq. 3.2).

Equations 3.1–3.4 enable the transformation of any original theoretical set of atomic charges corresponding to a well-defined value of n and of q_C° into a set corresponding to another value of n, without requiring the knowledge of the "inductive" formulas. In cyclohexane, for example, the net charges deduced from fully optimized STO-3G calculations are[3] $q_C = 2.76$, q_H(equatorial) $= 1.76$, and q_H(axial) $= -4.52$ me. These optimized Mulliken charges correspond to those calculated for the alkanes (Table 2.8), for which $n^{Mull} = 1.3325$ and $q_C^{\circ Mull} = -20.96$ me. The modified population analysis is now applied to deduce the charges corresponding to $n = -4.4122$ from that corresponding to $n = 1.3325$. First, applying Eq. 3.3, we obtain $q_C^\circ = -4.4122(-20.96/1.3325) = 69.40$ me, which is the modified reference net charge (i.e., that of the ethane-C atom). Next, it follows from Eq. 3.4 that $p = 30.12$ me. Finally, we obtain the modified net charges of cyclohexane from Eqs. 3.1 and 3.2, i.e., q_H(equatorial) $= -28.36$, q_H(axial) $= -34.64$, and $q_C = 63.00$ me. In the scale of relative units, the latter is q_C(rel. un.) $= 63.00/69.40 = 0.9078$.

It is easy to show that the modification expressed by Eqs. 3.1 and 3.2 is equivalent to simply changing the n value in the "inductive" equations. Indeed, the results of Table 2.4 can be adequately represented by means of general formulas, namely, $(3n + x)/3n$ (primary carbon atoms), $(2n + x)/3n$ (secondary C), $(n + x)/3n$ (tertiary C), and $-x/n$ (quaternary C), where the x's are numbers resulting from the σ^*'s of the groups from which the molecules are constructed. Moreover, all these formulas can be expressed as follows, in their most general form,

$$q_{\mathrm{C}} = \frac{q_{\mathrm{C}}^{\circ}(N_{\mathrm{CH}}n + x)}{3n} \quad \text{(charge units)}.$$

In this fashion, the general expression of Eq. 3.2 becomes (in charge units)

$$\frac{q_{\mathrm{C}}^{\circ}(N_{\mathrm{CH}}n + x)}{3n} = \frac{q_{\mathrm{C}}^{\circ\,\mathrm{Mull}}(N_{\mathrm{CH}}n^{\mathrm{Mull}} + x)}{3n^{\mathrm{Mull}}} + N_{\mathrm{CH}}p$$

which is an identity because of Eqs. 3.3 and 3.4.

It is now clear that the "inductive" formulas describing a selected group of alkanes are not only a summary of sets of theoretical results for atomic charges derived from different theoretical models, but also allow the study, within the framework of each theoretical approach, of the effect of changing the mode of partitioning CH overlap populations. We shall now examine the latter aspect in order to understand, at least in part, the origin of the wide discrepancies in the charge results given by the different theoretical methods.

3. NUMERICAL COMPARISONS OF ATOMIC CHARGES

On one hand, atomic charges derived from different theoretical approaches have a strong common point of resemblance, in that they reflect the customary inductive effects of alkyl groups; on the other, however, they differ numerically from one another depending upon the precise mode of calculation. We examine now the latter point, bearing in mind that our modification of the population analysis concerns only carbon-hydrogen overlap populations, retaining the usual half-and-half partitioning as far as carbon-carbon overlap terms are concerned, even when the charges on neighboring carbons are different. The charges of alkyl groups attached only to carbon atoms are, obviously, not affected by this modification, i.e., they are the same as in the Mulliken analysis. The results indicated in Table 3.1 reveal that the group charges predicted by the different methods are not nearly as dissimilar as one might expect from a superficial inspection of the individual atomic charges given in Tables 2.5–2.8. In fact, the comparison presented in Table 3.1 appears to be more significant than simple comparisons of carbon and hydrogen Mulliken net charges, since the latter are strongly dependent on the particular mode of partitioning carbon-hydrogen overlap populations.

In line with the above results, the net charge of the neopentane quaternary carbon atom is also "nearly the same" when calculated from different methods, sharply contrasting with the methane carbon atom, for example.

TABLE 3.1. Net Charges of Isolated Groups[a]

Molecule	Group	"Inductive"	EHMO	CNDO/2	INDO	PCILO	STO-3G	BO
		n	9.5	0.35	-2	0.525	1.3325	25
Propane	CH$_3$	$q_C^\circ/3n$	-12.5	-6.5	-8	-6.2	-5.12	-7
Butane	CH$_3$	$1.15 q_C^\circ/3n$	-15	(-8.3)	-13	(-8.0)	(-6.03)	-9
Isobutane	CH$_3$	$1.9 q_C^\circ/3n$	-24.3	-11.7	-19	-13.6	-9.96	(-13)
Neopentane	CH$_3$	$3 q_C^\circ/3n$	-34.3	-15.4	-23	-20.5	-15.73	(-20)
Propane	CH$_2$	$-2 q_C^\circ/3n$	+25	+13	+20	+12.4	+10.34	+17
Butane	CH$_2$	$-1.15 q_C^\circ/3n$	+14	(+8.3)	+13	(+8.0)	(+6.03)	+9
Isobutane	CH	$-5.7 q_C^\circ/3n$	+73	+35.1	+55	+40.8	+29.89	(+38)
Neopentane	C$_{quat}$	$-4 q_C^\circ/n$	+137	+61.5	+97	+82.0	+62.92	(+81)

[a] The results indicated in parentheses are calculated from the inductive formulas. All other values are deduced from the original literature[2].

TABLE 3.2. Population Analyses Modified for $n = -4.4122$ (me)

Molecule	Atom	Modified PCILO	"Inductive" from PCILO	STO–3G Modified	BO Modified
Methane	C	94.0	95.1	71.56	92.2
Ethane	C	92.6	92.2	69.40	89.4
Propane	C_{prim}	88.4	88.4	66.55	85.4
	C_{sec}	88.8	87.9	66.18	92.6
	H_{prim}	−32.3	−31.8	−23.89	−30.8
	H_{sec}	−38.2	−37.0	−27.92	−37.8
Butane	C_{prim}		89.2		89.4
	C_{sec}		84.8		86.6
	H_{prim}		−32.4		−32.8
	H_{sec}		−38.4		−38.8
Isobutane	C_{prim}	83.4	85.0	63.97	
	C_{tert}	85.6	84.4	63.48	
	H_{prim}	−33.2	−32.7	−24.62	
	H_{tert}	−44.8	−44.7	−33.65	
Neopentane	C_{prim}	83.0	82.0	61.70	
	C_{quat}	82.0	83.6	62.92	
	H	−34.1	−34.3	−25.81	

The appropriate p values are, in me, 34.32 (PCILO), 30.12 (optimized STO–3G), and 197.8 (BO). The nonmodified charges which were used for deriving these results are given in Table 2.5 (PCILO), 2.7 (BO), and 2.8 (STO–3G). The "inductive" charges were derived from Table 2.4, with $n = -4.4122$ and $q_C^\circ = 92.2$ me.

Typical *ab initio* results for q(quaternary C) are: 62.92 me[4] (STO–3G, optimized), 60.1 me[4] [GTO($6s3p|3s$)], 54 me[1] [from GTO($7s3p|3s$) results], and 81 me[1] (from Hoyland's BO results). Finally, from contracted GTO ($9s5p|5s$) calculations, $q(C_{quat})$ is estimated at ~ 90 me[4]. Now, it follows from the "inductive" equation $q(C_{quat}) = -4q_C^\circ/n$ that

$$q_C^\circ = \frac{-nq(C_{quat})}{4},$$

which means that *the ethane-C charge and, hence, all sets of carbon charges in alkanes deduced from different theoretical methods and modified for a common n value are not nearly as dissimilar (in going from one set to another) as their original counterparts*, because $q(C_{quat})$ and, thus, $nq(C_{quat})$ are nearly constant.

This theoretical conclusion can best be illustrated as follows, by transforming net charges derived from different theoretical methods into equivalent sets of results corresponding to any n value of our choice. In the following examples, n is taken at -4.4122 and the calculations are carried out on the basis of optimized STO–3G results as indicated in the example worked out in Section 2 ($p = 30.12$ me). Similar recalculations from PCILO[5] and BO[6] results are also reported (Table 3.2) for $n = -4.4122$. These modified charges

TABLE 3.3. A Comparison of STO–3G and GTO($6s3p|3s$) Results Modified for $n = 42.3$ with *ab initio* $7s3p|3s$ Results (electron units)

| Molecule | Atom | Net charge | | |
		Modified STO–3G	Modified GTO	$7s3p|3s$
Methane	C	−0.84	−0.95	−0.79
Ethane	C	−0.61	−0.70	−0.57
Propane	C_{prim}	−0.62	−0.70	−0.58
	C_{sec}	−0.39	−0.44	−0.38
Butane	C_{prim}	−0.61[a]	−0.70	−0.56
	C_{sec}	−0.39[a]	−0.45	−0.37
Isobutane	C_{prim}	−0.62	−0.70	−0.55
	C_{tert}	−0.16	−0.18	−0.18

[a] Calculated using the "inductive" formulas for the original STO–3G and GTO charges.

TABLE 3.4. A Comparison of the Ratio $q_C^\circ/q_C^\circ(Av)$ Calculated for a Common n Value

Method	$q_C^\circ/q_C^\circ(Av)$	Method	$q_C^\circ/q_C^\circ(Av)$
INDO	1.5	STO–3G standard	1.1
CNDO/2	0.94	STO–3G optimized	0.68
PCILO	0.90	$6s3p|3s$	0.69
RCNDO	1.0	$7s3p|3s$	0.58
PPP-type	0.73	BO	0.87
EHMO	1.6		

Calculated from the original charge distributions: CNDO/2 and PCILO, Ref. 5; INDO and RCNDO, Ref. 8; PPP-type, Ref. 9; EHMO, Ref. 10; STO–3G and $6s3p|3s$, Ref. 4; $7s3p|3s$, Ref. 7; BO, Ref. 6.

turn out to be similar in magnitude to the corresponding STO–3G charges, quite unlike their original counterparts.

Another example indicating that very dissimilar original sets of charge distributions generate sets of similar magnitude after correction to a common n value is presented in Table 3.3. Here, the STO–3G results of Table 2.8 and the GTO($6s3p|3s$) results of Table 2.7 are modified for $n = 42.3$, which is the n value corresponding to Leroy's GTO($7s3p|3s$) calculations[7] given in Table 2.7.

The similarity of the charges derived from a variety of methods after modification to a common n value is demonstrated in Table 3.4. This comparison is made as follows. First, we calculate the q_C° value corresponding to each theoretical method rescaled to a common n value, chosen arbitrarily, by means of Eq. 3.3. Next, the average $q_C^\circ(Av)$ is calculated from these data and is chosen as a common arbitrary reference. Finally, the ratios $q_C^\circ/q_C^\circ(Av)$

are calculated. These ratios indicate that the modified charges given by each selected method are $q_C^\circ/q_C^\circ(Av)$ times those of the corresponding charges of the standard of reference. The fact that for both semi-empirical and *ab initio* calculations all the $q_C^\circ/q_C^\circ(Av)$ values lie in a narrow range indicates that net charges modified for a common n value are not as dissimilar as their original counterparts and confirms the prediction inferred from the inductive theory. We still do not know, however, which is the physically correct n.

4. THE "MOST EVEN ELECTRON DISTRIBUTION"

The arguments developed in Section 1 of this Chapter demonstrate that the use of atomic charges corresponding to the "proper" n value is crucial in any study of property-charge relationships involving carbon atoms. Moreover, by means of Eqs. 3.1–3.4, it is now easy to modify theoretical sets of charges in order to obtain charges corresponding to any n value of our choice. The "true" value of n, however, still remains to be defined. In principle, the appropriate value of n should be determined from an examination of property-charge relationships involving experimental results. Fortunately, this turns out to be relatively simple provided that adequate sets of theoretical and experimental results are available. There is, however, a particular value of n which merits attention: that corresponding, in a way, to a condition of "most even electron distribution" [2].

This concept derives from customary intuitive views: the electron-attracting power of otherwise similar atoms decreases as their electron populations increase, thus opposing charge separation. For a set of alkane molecules, each of which contains two different carbon atoms with net charges q_i and q_j, a possible formulation can be derived from the condition that the sum $\Sigma(q_i - q_j)^2$ (over the set) be minimum, i.e., that the various carbon atoms of the set be as similar as possible to one another. Using the "inductive" equations of Table 2.4, it is found that the equation

$$\frac{\partial \Sigma(q_i - q_j)^2}{\partial n} = 0$$

is satisfied by $n = -4.4$. Alternatively, using the original optimized STO–3G Mulliken charges of Table 2.8, one can calculate the p value which generates a set of charges as similar as possible to a constant, k, from the condition

$$\Sigma(q_C^{Mull} + N_{CH}p - k)^2 \text{ minimum,}$$

which is expressed on the basis of Eq. 3.2. This calculation yields $p \simeq 30$ me and, from Eqs. 3.3 and 3.4, $n = -4.4$.

Admittedly, these estimates of n are open to criticism on the grounds of some (unavoidable) arbitrariness introduced with the sampling of the molecules used for these calculations. Nevertheless, it remains that the n value best describing the "most even electron distribution" in alkanes lies somewhere near -4.4. This n value now appears to correspond to real physical situations. Indeed, comparisons of experimental results with carbon net charges indicate $n = -4.4122$ (from C–13 nuclear magnetic resonance shifts[2]), $n = -4.4083$ (from ionization potentials[11]), and $n = -4.41$ (from energies of atomization[12]). Consequently, it can be assumed with a reasonable degree of confidence that $n \simeq -4.4$ best describes the charge distributions in saturated hydrocarbons. Finally, it is interesting to point out that SCF–Xα–SW atomic charges[13], which do *not* derive from a Mulliken population analysis, correspond to $n = -4.429$.

5. ON THE PRECISION OF THEORETICAL CHARGES

The satisfactory agreement between "inductive" and quantum mechanical charge distributions may suggest that, after all, it does not really matter which method is used for calculating charges because all results are alike in that they reflect the "inductive" equations, except for the corresponding n and q_C° values, and, moreover, because the transformation of the initial sets of results into sets corresponding to the appropriate n value is easy. Unfortunately, this is not quite the case as the choice of the theoretical method turns out to be most important as regards the precision and the internal consistency of its results. Indeed, at the level $n \simeq -4.4$, the whole range of carbon charges is covered by values which differ relatively little from one another, namely, of the order of $\sim 13\%$ between the extremes. This is clear when considering the methane carbon atom ($q_C = 1.0311$ relative units) and the neopentane quaternary carbon atom ($q_C = 0.9066$ rel. un.). Under these circumstances, even a 1% difference in charge is significant as it is larger than, for example, the differences between the primary and secondary carbon atoms in propane, or the primary and tertiary carbon atoms in isobutane (Table 3.5).

In terms of fully optimized STO–3G charges, for example, modified for $n = -4.4122$, the carbon net charges are 71.56 me in methane and 69.40 me in ethane, which represents a difference of only 2.16 me between these two C atoms. Had each C atom been miscalculated by only 1 me in the original calculations, the corrected charges and, hence, the relative charges deduced therefrom, would have been seriously affected. This arises from the nature of the correction $q_C = q_C^{Mull} + N_{CH}p$ which carries any error in the evaluation of q_C^{Mull} into the final value q_C. While in some cases this error may be termed minor (in a relative sense) in the q_C^{Mull} set of results, the same absolute error may be too far large in comparisons involving the corrected values, q_C.

Considering the extreme smallness of acceptable errors, one may fear to be unable to reproduce sufficiently accurate charge distributions for use in comparisons involving experimental properties which are highly sensitive to charges. Fortunately, the "inductive" charges are not only in best agreement with the most accurate set of theoretical charges (as shown in Table 2.8) but

TABLE 3.5. Charge Distributions of Alkanes for $n = -4.4122$ in Relative Units

Molecule	Atom	Net Charge
Methane	C	1.0311
Ethane	C	1.0000
Propane	C_{prim}	0.9584
	C_{sec}	0.9537
	H_{prim}[†]	−0.3447
	H_{sec}	−0.4013
Butane	C_{prim}	0.9675
	C_{sec}	0.9198
	H_{prim}	−0.3515
	H_{sec}	−0.4164
Pentane	C_{centr}	0.8782
	H_{centr}	−0.4278
Isobutane	C_{prim}	0.9222
	C_{tert}	0.9151
	H_{prim}[†]	−0.3552
	H_{tert}	−0.4844
Neopentane	C_{prim}	0.8889
	C_{quat}	0.9066
	H	−0.3719

[†] Weighted average of nonequivalent H atoms.

also yield the best correlations when used in comparisons with experimental data, with a precision estimated at $\sim 0.13\%$ (Chapters 4, 6). The very fact that these diametrically different approaches are virtually in perfect agreement when compared to one another and both are most successful when used in comparisons with experiment naturally points at STO–3G calculations which can and should be used for the study of property-charge relationships provided, of course, that full geometry and ζ exponent optimizations are carried out. Larger basis set calculations would evidently be preferable for various theoretical reasons (other than those linked to charge distributions) but the major difficulty would then rest with the practical feasibility of all required optimizations which would ensure the internal consistency of an adequate set of charge results.

6. INDUCTIVE EFFECTS AND CHARGE DISTRIBUTIONS

The inductive effects of alkyl groups, or, more precisely, some aspects linked to intramolecular charge transfers, have been used as a useful guideline and tool for discussing theoretical charge distributions, with the result that atomic charges derived from different theoretical models are unified in a more general picture covering also the effects of changing the mode of partitioning overlap populations. As such, however, inductive effects have not been discussed in their own right, from the customary "chemical" point of view. This discussion is presented here[14], with reference to the "final" charge results defined as follows.

As regards n, its value ($n = -4.4122$) is selected on the basis of carbon–13 NMR shift results[2], as described in Chapter 4.4. This n value indicates a C^+—H^- polarity in alkanes, a result which is in line with the view that hydrogen is certainly more electronegative than carbon, as Mulliken and Roothaan have pointed out[15]. Since then, the C^+—H^- polarity has been advocated on various occasions, e.g., by Julg[16], Bader[17], Jug[18], Wiberg[19], and in localized MO theory[20]. On the other hand, it is also difficult to extract q_C° from theoretical calculations only, because we do not know for sure what level of calculation fulfills the minimum requirements for a reliable result. Optimizations, namely those of the ζ exponents, and increasingly larger basis set descriptions are known to reduce charge separations and, hence, q_C° (Table 3.4). For example, exponent optimization reduces the STO–3G value of q_C° from ~ 114 me to 69.40 me (for $n = -4.4122$), whereas nonoptimized 6–31G calculations give ~ 58 me. We adopt in the following $q_C^\circ = 35.1$ me, which is the value indicated by the study of the energies of atomization[12] (Chapter 6.4) and use this result and the relative charges of Table 3.5 for discussing the numerical examples given below. Of course, the validity of the forthcoming arguments does not depend on the actual choice of q_C°.

The C^+—H^- polarity corresponds to the intuitive picture suggested by the group charges of Table 3.1. These charges, which are not affected by the modification concerning the partitioning of CH overlap terms, indicate that the retention of electrons in a CH_x group increases as the number, x, of H atoms in the group becomes larger, and vice versa. So, for example, when compared to a CH_2 group, a CH group loses about twice as much charge to each adjacent CH_3 group. Of course, the charge transfer occurs in favor of the hydrogen-richer group. This suggests that, in a way, the hydrogen atoms "retain" electrons in the groups, in qualitative agreement with a scheme depicting the hydrogens as electron attractors, i.e., with the C^+—H^- polarity. This qualitative picture of electron retention by the hydrogen atoms in CH_x groups is the central argument from which the inductive effects of the alkyl groups can be rationalized. (The quantitative energetic aspects directly related to the concept of hydrogen atoms acting as "electron reservoirs" are given in Chapter 5.9.)

The following rationale [14] clarifies various aspects relating inductive effects to the C^+—H^- polarity. Isobutane is chosen as an example for illustrating the effect of α-methyl substitution on a "central" carbon atom (C_{tert}, in this case) and on a "central" CH_x group (C_{tert}—H_{tert}, in this example). This molecule may be described as in 1, using the charges of

Table 3.5 (for $n = -4.4122$) and $q_C^\circ = 35.1$ me, or, alternatively, as in **2**—a representation which is somewhat misleading.

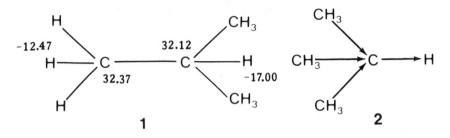

1   **2**

In **2** the arrow $C \rightarrow H$ suggests correctly that the tertiary hydrogen atom gains electronic charge. The arrows $CH_3 \rightarrow C_{tert}$ suggest that the central carbon atom gains electrons as the number of CH_3 groups replacing H is increased. This is in line with common views and is, indeed, reflected by the population analyses for $n = -4.4122$ (e.g., propane-C_{sec}, $q_C = 33.48$ me; isobutane-C_{tert}, 32.12 me; neopentane-C_{quat}, 31.82 me). Nevertheless, the arrow $CH_3 \rightarrow C$ is misleading in another respect, in that it may suggest that the methyl group loses electronic charge to the rest of the molecule, whereas all calculations (modified or nonmodified) indicate the reverse, i.e., a net negative charge for each methyl group when attached to a better electron-donor (iso-C_3H_7, in this example). Reversal of the arrow ($CH_3 \leftarrow C$), intending to account for $q(CH_3)$ negative, is no better because then it would seem that the central atom becomes increasingly positive as the number of methyl substituents is increased. This is false for $n = -4.4122$. (Any calculation made with theoretical methods corresponding to $n > 0$ indicates, on the contrary, that the central C atoms become increasingly positive as the number of methyl substituents is increased. This is certainly a good source of confusion.) The apparent conflict between the facts (*i*) that the central carbon atom gains electrons with increasing number of methyl substituents and (*ii*) that the methyl groups themselves withdraw electronic charge is easily resolved by observing that although each methyl group pulls electrons from the iso-C_3H_7 group [$q(CH_3) = -5.04$ me], it attracts less electrons than the hydrogen atom which it has replaced [$q_H(sec)$ in propane $= -14.09$ me]. This, of course, means that hydrogen is a better attractor than CH_3, which is reflected in methane, $(CH_3)^+$—H^-, i.e., by the C^+—H^- polarity. Moreover, as discussed above, the CH group retains electrons less efficiently than CH_3, thus developing a $(CH)^+$—$(CH_3)^-$ polarity, which means that the partial charge transfer necessarily takes place from the iso-C_3H_7 group towards the CH_3 group. Hence, iso-C_3H_7 appears to be a better electron donor than CH_3. Similar arguments also apply to other molecules. In short, *the central carbon atom becomes increasingly negative (i.e., less positive) as the number of α-methyl substituents is increased not because methyl itself pushes electrons toward the central carbon atom but because CH_3*

withdraws less electrons than the hydrogen atom for which it has been substituted. As a whole, we recognize here many of the familiar pictures and aspects of the customary interpretation of the inductive effects of alkyl groups which are well supported by a vast body of chemical information. The point is that the present rationale is a product of detailed theoretical charge analyses, namely, of the results indicating (i) a C^+—H^- polarity (describing hydrogen atoms as "electron reservoirs") and (ii) minimal changes in electron populations at the carbon atoms, (i) being a consequence of (ii).

7. CONCLUSIONS

For a selected set of alkanes, atomic charges expressed as $f(n, q_C^\circ)$ represent a summary of results given by a host of different theoretical approaches, namely, by *ab initio* calculations differing in their basis set descriptions. The parameter n controls the relative ordering of the carbon net charges and its value is of crucial importance in any comparison involving them in the study of property-charge relationships. Now we have seen that within the framework of any individual LCAO method the value of n is determined by the mode of partitioning overlap populations and that the "inductive" charges $f(n, q_C^\circ)$ retain their full validity in describing sets of modified carbon charges, $q_C = q_C^{Mull} + N_{CH}p$, with respect to sets of Mulliken charges. Indeed, n and p are related as follows to one another

$$p = \frac{q_C^{\circ\,Mull}(n - n^{Mull})}{3n^{Mull}}$$

(from Eqs. 3.3 and 3.4). Consequently, we can now transform any original (e.g., Mulliken) set of charges corresponding to definite values of n and q_C° into a set corresponding to another value of n without having recourse to the "inductive" formulas $f(n, q_C^\circ)$. For example, the Mulliken net charges of the carbon atoms in methane and ethane are, from optimized STO–3G calculations, -48.92 and -20.96 me, respectively, giving $48.92/20.96 = 4(n + 1)/3n$ and $n = 1.3325$. Hence, the p value transforming this type of STO–3G results into those for $n = -4.4122$ is 30.12 me, which is used for deducing the corrected q_C and q_H charges from their Mulliken counterparts (Eqs. 3.1, 3.2). A remarkable result is the following: when atomic charges in alkanes deduced from different theoretical methods (say, methods A, B, etc.) are modified for a common n value, the corrected sets of results (corresponding to methods A, B, etc.) are relatively similar to one another, quite unlike their original, uncorrected, counterparts. This result certainly represents some relief in questions regarding the validity of charge distributions predicted by different methods, namely, with the use of different basis sets in *ab initio* calculations.

Charge distributions in alkanes corresponding to a "most even electron distribution" are obtained with $n \simeq -4.4$ and describe, in a way, carbon atoms allowing "most reluctantly" changes in their electron populations. The inductive effects of alkyl groups are rooted in the C^+—H^- polarity, a basic result which is a consequence of $n \simeq -4.4$: the "retention" of negative charge by the hydrogen atoms is coherent with, and reflected by, the net electron transfer from the CH_x groups containing fewer H atoms toward the CH_x groups possessing more H atoms (e.g., from CH_2 toward CH_3). This role of the hydrogen atoms acting as "electron reservoirs" provides a simple rationale for the inductive effects exhibited by the alkyl groups, in line with all known facets deduced from experimental evidence and with theoretical charge distributions.

So far, the present approach has only dealt with saturated hydrocarbons. With these compounds there is only one partitioning problem, that of the carbon-hydrogen overlap term. Introduction of heteroatoms (or sp^2 and sp carbon atoms) involves the knowledge of other modes of partitioning overlap populations. This remark alone defines the problem arising with molecules other than saturated hydrocarbons. With amines, for example, it would be necessary to consider the mode of partitioning C—N and N—H overlap terms; otherwise it is likely that charge distributions would be obtained which are no better in quality than the uncorrected ones for the hydrocarbons, i.e., charges which in all likelihood could not be used in direct comparisons with observed properties. Clearly, the most logical approach to the heteroatom problem would involve applying appropriate "p-corrections" similar to those met in the study of hydrocarbons. There are, however, situations which can be studied without applying the "p-corrections". For example, in the series R—NH_2 (R = alkyl) the charges on nitrogen (or the NH_2 group) could be used without correction, for comparative purposes, because the "p-corrections" for the C—N and N—H overlap terms appear in all cases the same number of times. In general, *uncorrected sets of charge distributions can, in principle, be used for comparative purposes within given series of molecules when the sum of the "p-correction" terms remains the same throughout the series.* (This point is illustrated by the group charges indicated in Table 2.2 and 3.1.) Advantage is taken of this possibility in studies involving oxygen atoms[21] (Chapter 4). In a series of dialkylethers, for example, the Mulliken net charges of oxygen atoms are compared with their nuclear magnetic resonance shifts; a correlation of this sort involves primarily the *difference* in charge at oxygen in going from one compound to the other and not absolute values of net charges, whose theoretical evaluation would require the "p-correction" for the two ether C—O bonds. In the present case, adequate estimates of the distribution of electronic charges can nonetheless be gained from charge normalizations based on the detailed knowledge acquired so far about the charges in alkyl groups (Chapter 4.6). While a number of challenging problems can be solved in a similar manner, at least approximately, it remains that a general treatment of the way overlap populations should be partitioned still awaits further investigations. Valuable guidelines are offered by nuclear magnetic resonance studies, which are taken on in the following chapter.

REFERENCES

1. S. Fliszár, G. Kean, and R. Macaulay, *J. Am. Chem. Soc.*, **96**, 4353 (1974).
2. S. Fliszár, A. Goursot, and H. Dugas, *J. Am. Chem. Soc.*, **96**, 4358 (1974).
3. R. Roberge and S. Fliszár, *Can. J. Chem.*, **53**, 2400 (1975).
4. G. Kean and S. Fliszár, *Can. J. Chem.*, **52**, 2772 (1974).
5. S. Fliszár and J. Sygusch, *Can. J. Chem.*, **51**, 991 (1973).
6. J.R. Hoyland, *J. Chem. Phys.*, **50**, 473 (1969).
7. J.M. André, P. Degand, and G. Leroy, *Bull. Soc. Chim. Belg.*, **80**, 585 (1971).
8. D.R. Salahub and C. Sándorfy, *Theor. Chim. Acta*, **20**, 227 (1971).
9. S. Katagiri and C. Sándorfy, *Theor. Chim. Acta*, **4**, 203 (1966).
10. R. Hoffmann, *J. Chem. Phys.*, **39**, 1397 (1963).
11. H. Henry and S. Fliszár, *Can. J. Chem.*, **52**, 3799 (1974).
12. S. Fliszár, *J. Am. Chem. Soc.*, **102**, 6946 (1980).
13. A.E. Foti, V.H. Smith, and S. Fliszár, *J. Mol. Struct.*, **68**, 227 (1980).
14. S. Fliszár, *Can. J. Chem.*, **54**, 2839 (1976).
15. R.S. Mulliken and C.C. Roothaan, *Chem. Rev.*, **41**, 219 (1947).
16. A. Julg, *J. Chim. Phys.*, **53**, 548 (1956).
17. R.F.W. Bader and H.J.T. Preston, *Theor. Chim. Acta*, **17**, 384 (1970).
18. K. Jug, *Theor. Chim. Acta*, **31**, 63 (1973); *ibid.*, **39**, 301 (1975).
19. K.B. Wiberg and J.J. Wendoloski, *J. Comput. Chem.*, **2**, 53 (1981).
20. M.S. Gordon and W. England, *J. Am. Chem. Soc.*, **94**, 5168 (1972); R.H. Pritchard and C.W. Kern, *J. Am. Chem. Soc.*, **91**, 1631 (1969).
21. M.-T. Béraldin, E. Vauthier, and S. Fliszár, *Can. J. Chem.*, **60**, 106 (1982).

CHAPTER 4

Charge Analyses Involving
Nuclear Magnetic Resonance Shifts

1. INTRODUCTION

The critical examination of charge analyses derived in the spirit of a generalized Mulliken scheme (Eqs. 1.3–1.5) indicates that the results obtained from different theoretical models are in general agreement. The validity of this unifying conclusion has been established in a detailed fashion only for the saturated hydrocarbons but it is clear that departures from the usual halving of overlap populations are to be considered in any case involving heteronuclear overlap partners. Unfortunately, the practical extension of these views to other classes of compounds is not yet feasible because of the lack of precise guidelines providing a general recipe for the appropriate partitioning of overlap populations. Therefore, it becomes necessary in some cases to use indirect methods for obtaining charge analyses by putting common sense to work as a temporary replacement for mathematical formulations.

In the effort to maximize our knowledge concerning carbon and hydrogen atomic charges of organic molecules, leading eventually to atomic charges of heteroatoms from charge normalization, we meet with the formidable task of calculating large collections of molecules keeping in mind the requirements for accuracy described in Chapter 3.5. Namely, in the simplest possible approach, this would imply a large number of *ab initio* calculations, involving full optimization of all geometrical parameters and of all scale factors in order to produce sets of sufficiently accurate Mulliken charges, which could be used as a starting point for the derivation of "final" charges reflecting an appropriate partitioning of overlap populations. While, of course, such an approach remains always possible in principle, we prefer to explore an alternative route in terms of carefully established empirical correlations between atomic charges and nuclear magnetic resonance shifts, and to verify their validity by carrying out the appropriate spot tests by means of accurate calculations of representative molecules. In this fashion we gain access to numerical charge results for a considerable number of molecules with an accuracy which can be evaluated with a reasonable degree of confidence. The loss in theoretical elegance is largely compensated by the practical feasibility of this type of approach.

Unfortunately, charge–NMR shift relationships are not free from conceptual difficulties and have drawbacks of their own[1-3]. It is therefore important to be fully aware both of the possibilities offered by such correlations and of the limitations which restrict their utilization.

2. RELATIONSHIPS BETWEEN NUCLEAR MAGNETIC RESONANCE SHIFTS AND ATOMIC CHARGES

Before proceeding with detailed analyses of charge–NMR shift correlations, the examination of a few basic aspects regarding this type of correlation is in order. The main conceptual difficulty stems from the fact that the attempts at correlating NMR shifts with atomic electron populations are rooted in intuition rather than being based on a formalism explicitly featuring the role of local charges in governing shielding constants. This situation paves the way to criticisms which are countered, in essence, by a significant number of "good" charge–shift relationships, although the reason(s) for, or validity of, this type of results remains always difficult to assess. Fortunately, we can take advantage of an indirect way of assessing the merits of charge-shift correlations by examining the average diamagnetic and paramagnetic contributions, σ^d and σ^p, respectively, to the total average magnetic shielding

$$\sigma = \sigma^d + \sigma^p \qquad (4.1)$$

The general theory of magnetic shielding, the result of a number of theoretical studies[4], is now clearly described in a rigorous manner. The rigorous theory, however, is practically unmanageable for any compound of interest, and the manageable approaches, from the point of view of computational feasibility, suffer from so many approximations that they lose their rigor. Reasonable compromises between theoretical rigor and practical feasibility have been offered in four types of approaches, namely (*i*) the coupled Hartree–Fock method[5,6], (*ii*) the finite perturbations method[7], (*iii*) Pople's classical perturbation theory[8], and (*iv*) the variation method[9,10]. *Ab initio* calculations carried out within the framework of the coupled Hartree–Fock method yield results which depend on the choice of the origin of the molecular axes[11]. In order to avoid this drawback, Ditchfield[12] has introduced gauge invariant atomic orbitals (GIAO). The GIAO's were also used in applications of Pople's classical method at the CNDO/S and INDO levels[13] and, in the variation method[14], at the Extended Hückel and INDO levels[15]. The σ^d and σ^p results used in the forthcoming discussion were derived using the formalism given by Vauthier, Tonnard and Odiot[16]. This approach is rooted in Pople's finite perturbation theory[7]. It involves the INDO approximations[17] on a GIAO basis and London's approximation. Moreover, it satisfies the Hermitian requirement for the first-order perturbation matrix reflecting the effect of an applied external magnetic field. The latter condition results in a significant improvement of calculated ^{13}C magnetic shieldings, the average precision being of the order of ~ 5 ppm[16]. The point is that this formalism for σ permits a separation into mono-, di-, and triatomic contributions, thus revealing the relative importance of "local" and "distant" electron densities on the magnetic shielding of a given nucleus. In this manner, it becomes possible to gain a reasonable estimate about the chances that chemical shifts do, indeed, depend primarily on local electronic populations, at least in series of closely related compounds[16,18]. The most detailed results are those derived for ethylenic and acetylenic sp^2 and sp carbon atoms, respectively.

To begin with, it appears that the local diamagnetic contribution to the magnetic shielding is practically the same for all sp^3, sp^2 and sp carbon nuclei (57.85 $\pm$ 0.6 ppm). Moreover, the results for sp^2 carbons indicate that the *total* diamagnetic part (including all contributions from distant atoms) *plus* the paramagnetic part due to the distant atoms is nearly constant (82.7 ppm), within ~ 0.4 ppm. The gap between this sum and the total magnetic shielding represents the paramagnetic contribution excluding that of distant atoms, i.e., the local paramagnetic shielding plus the paramagnetic part contributed by the neighbors of the nucleus under study. It is this gap which reflects the total variation in magnetic shielding (or, at least, its major part by far) for a given nucleus in a series of closely related compounds; it is now at the center of our attention. The effects of the neighboring atoms which are included in this paramagnetic shielding are reported in Table 4.1. The results reflect the smallness of these effects.

For ethylenic and acetylenic carbon atoms, one can consider the neighbors' contributions as being constant, or nearly so (within ~ 1.5 ppm), and the corresponding uncertainty introduced by assuming constant neighbors' contributions for sp^3 carbon atoms does probably not exceed ~ 0.3 ppm. As a consequence, in a series of closely related compounds, *the variations of the local paramagnetic shielding appear to represent the largest part, by far, of the total changes in shielding* experienced by a given nucleus due to structural changes, e.g., by sp^2 carbons in a series of ethylenes. Therefore, within the precision of the present type of analysis, it seems quite reasonable to anticipate correlations between nuclear magnetic resonance shifts and atomic charges which, of course, are strictly local properties. Following this analysis of the individual nonlocal effects revealing, namely, the small participation of tricentric integrals involving distant atoms, the practical validity of

TABLE 4.1. Paramagnetic Shielding Contributed by Neighboring Atoms (ppm)[18]

Molecule (Atom*)	Shielding	Molecule (Atom*)	Shielding
C^*H_4	0.17	$(CH_3)_2C=C^*HCH_3$	0.38
$CH_3C^*H_3$	−0.11	$(CH_3)_2C^*=CH_2$	−1.15
$CH_2=C^*H_2$	−1.21	$(CH_3)_2C^*=CHCH_3$	1.22
$CH_3CH=C^*H_2$	0.70	$(CH_3)_2C^*=C(CH_3)_2$	1.52
$(CH_3)_2C=C^*H_2$	1.52	$CH\equiv C^*H$	1.65
$CH_3C^*H=CH_2$	−2.15	$CH_3C\equiv C^*H$	3.79
$CH_3C^*H=CHCH_3$ cis	0.62	$CH_3C^*\equiv CH$	1.15
$CH_3C^*H=CHCH_3$ trans	0.67	$CH_3C^*\equiv CCH_3$	3.09

These results were calculated from those indicated in Ref. 16 and represent $\sigma^p(KK) + \sigma_0^p(MK)$, as defined in this reference. The local paramagnetic shielding discussed in the text is $\sigma^p(M)$ (Eq. 9 of Ref. 16).

charge–shift relationships rests largely with cancellation effects of a number of terms which, to begin with, are small or relatively constant. The importance of the nonlocal contributions is further reduced with the selection of a scale tailored for comparisons between atoms of the same type, with reference to an appropriately chosen member of that series. Finally, it has been pointed out[16] that the small participation of distant atoms to the magnetic shielding of carbon nuclei may justify the smallness of solvent effects on ^{13}C shifts and the solvaton model[19] used by Webb[20].

The justification outlined here for carbon–13 charge–shift relationships applies also to oxygen and nitrogen atoms[21]. For hydrogen atoms, however, the situation is entirely different. The most striking aspect is the weight of the three-center integrals in the calculation of their magnetic shielding. Indeed, the evaluation of proton shieldings yields erratic results if three-center integrals are not retained, whereas their inclusion restores the correct relative ordering of the resonance shifts[16]: the marked response to distant structural modifications, including solvent effects, may be regarded as a reflection of the important contributions arising from three-center integrals.

So far we have learned that, in certain series of closely related compounds, it is the local paramagnetic shielding which governs the changes in total shielding, i.e.,

$$\Delta\sigma_{total} \simeq \Delta\sigma_{local}^p \qquad (4.2)$$

and, hence, that under these circumstances it may well be justified to expect correlations between nuclear magnetic resonance shifts and local electron populations. It remains, however, that charge–shift correlations are essentially empirical in nature: while the definition of "closely related compounds" may be linked to the approximate validity of Eq. 4.2, the practical answer stems ultimately from the actual examination of shift vs. charge results. This is to say that satisfactory correlations found for given series of compounds should not be used to lend unintended support for the unwarranted view that they are necessarily valid in all systems. For example, a good correlation between ^{17}O NMR shifts and atomic charges in dialkylethers does not imply its validity in acetals, in which effects other than simple inductive effects play a significant role in determining electron distributions at the s and p levels.

Of course, the validity of any correlation between experimental chemical shifts and calculated atomic charges clearly depends on the validity of the theoretical charge analysis. Interestingly enough, little consideration is usually given to the latter point and a number of different theoretical methods, both semi-empirical and *ab initio*, are used almost indiscriminately for calculating charges for use in charge *vs.* shift correlations. In the forthcoming analyses, extensive use is made of the results described in Chapters 1–3, in particular those related to the mode of partitioning overlap populations in heteronuclear situations. With these premises, we can now examine the trends observed in charge–shift correlations.

For convenience, we express the chemical shifts, δ, of carbon–13 in parts per million (ppm), with reference to tetramethylsilane (TMS). In this scale convention, positive δ values correspond to downfield shifts. In writing a linear relationship

$$\delta = aq + b \qquad (4.3)$$

with net atomic charges, we keep in mind that $q < 0$ represents an excess negative charge (Eq. 1.2) and that q becomes more negative as the corresponding electron population increases. Hence, a positive slope a indicates that an increase of electronic charge at an atom results in an upfield shift, reflected by a lowering of δ. Conversely, a negative slope a indicates that an increase in local electron population (more negative q) results in a downfield shift. Both positive and negative slopes are met in applications of Eq. 4.3. This is a problem well worth looking into.

3. CHARGE–SHIFT RELATIONSHIPS INVOLVING sp^2 CARBON ATOMS

Let us first examine the probably most quoted plot, that of the familiar Spiesecke and Schneider work[22] relating the ^{13}C NMR shifts of tropylium ion, benzene, cyclopentadienyl anion, and cyclooctatetraene dianon to the corresponding carbon atomic charges. The latter were deduced by assuming the local π-electron density to be known from the number of π-electrons and the number of carbon atoms over which the π-cloud is distributed. The estimated shift, ~ 160 ppm per electron, has become an almost unerasable part of our grammar. The linear correlation between ^{13}C chemical shifts and π-charge density was later extended to 2π-electron systems[23-25], as well as to the 10π cyclononatetraene anion[26]. A plot of this correlation for the whole series was presented by Olah and Mateescu[23] who used, where appropriate,

TABLE 4.2. Mulliken Orbital Populations of Selected Aromatic Hydrocarbons (electron units)

Compound	Orbital Population			
	$1s$	$2s$	$2p_x + 2p_y$	$2p_z$
1 Cyclopropenium cation, $C_3H_3^+$	1.99232	1.18153	2.04962	0.66667
2 Cycloheptatriene cation, $C_7H_7^+$	1.99178	1.15295	1.97580	0.85714
3 Benzene, C_6H_6	1.99178	1.13282	1.92281	1.00000
4 Cyclononatetraenide anion, $C_9H_9^-$	1.99199	1.11693	1.88599	1.11111
5 Cyclopentadienide anion, $C_5H_5^-$	1.99218	1.11388	1.86741	1.20000
6 Cyclooctatetraenide dianion, $C_8H_8^=$	1.99217	1.10775	1.82507	1.25000

simple Hückel molecular orbital theory for deducing charge distributions. At a quite different level of approximation, this class of compounds was also investigated by means of STO–3G calculations involving a detailed optimization of all the geometrical and ζ exponent parameters[18]. The Mulliken orbital populations are indicated in Table 4.2. The corresponding net atomic charges and chemical shifts (Table 4.3) yield the correlation presented in Figure 4.1. In spite of some scatter about the correlation line, a point which is discussed further below, it appears that Eq. 4.3 is reasonably well satisfied with the use of total $(\sigma + \pi)$ net atomic charges, with $a \simeq 300$ ppm/electron.

A closely related example concerns the *para* carbon atoms of substituted benzenes. The Mulliken net charges, at the STO–3G level, were calculated by Hehre, Taft, and Topsom[3] who presented a most instructive study of these compounds, including detailed considerations of π- and σ-charge density changes in terms of inductive and conjugative substituent parameters; the concurrent use of substituent parameters and concepts of electron disturbance in monosubstituted benzenes appears to be well supported at the level of STO–3G theory. The charge results (relative to 1, 5 *viz.* 6 electron for π, σ, and total net charges, respectively) are indicated in Table 4.4,

TABLE 4.3. Carbon Net Charges and NMR Shifts of Selected Aromatic Hydrocarbons (me, *viz.* ppm from TMS)

Molecule	q_σ	q_π	q_{tot}	δ
1 $C_3H_3^+$	−223.5	333.3	109.8	176.8
2 $C_7H_7^+$	−120.5	142.9	22.4	155.4
3 C_6H_6	−47.4	0	−47.4	128.7
4 $C_9H_9^-$	5.1	−111.1	−106.0	108.8
5 $C_5H_5^-$	26.5	−200.0	−173.5	102.1
6 $C_8H_8^=$	75.0	−250.0	−175.0	85.3

Results calculated from Table 4.2, relative to 1, 5 and 6 electrons, respectively, for the π, σ and total net charges. A negative sign indicates an increase in electron population.

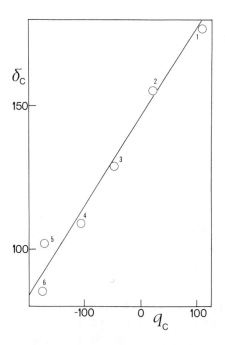

FIGURE 4.1. Carbon–13 chemical shifts of the aromatic compounds indicated in Table 4.3 *vs.* total $(\sigma + \pi)$ net charges (ppm from TMS, viz. 10^{-3} electron units). Reproduced from Ref. 18.

together with the corresponding NMR shifts. The correlation between the total net charges and the chemical shifts is shown in Figure 4.2 for the *para* carbons, with $a \simeq 384$ ppm/electron. A similar study on *meta* carbons, while giving results of the same type[3], is perhaps somewhat less conclusive because of the very limited range of variation of the *meta* carbon NMR shifts $(\sim 1.5$ ppm) and the difficulty of obtaining sufficiently accurate charge results at the "standard" STO–3G level. It remains, however, that the major conclusions drawn here and further below for the *para* carbons apply to the *meta* carbons as well.

TABLE 4.4. Charge Analysis and NMR Shifts of *para* Carbon Atoms in Substituted Benzenes

Substituent	q_π (me)	q_σ (me)	δ (ppm)
1 NH_2	-46	-35	119.2
2 OH	-39	-39	120.8
3 F	-21	-50	124.3
4 CH_3	-12	-56	125.6
5 H	0	-63	128.7
6 CN	28	-78	130.1
7 NO_2	43	-87	134.7

The Mulliken net charges are those given in Ref. 3. The shift results are taken from J.B. Stothers, "Carbon–13 NMR Spectroscopy", Academic Press, New York, NY, 1972.

FIGURE 4.2. Correlation between carbon–13 NMR shifts and total $(\sigma + \pi)$ net charges of *para* carbon atoms in substituted benzenes, from the results indicated in Table 4.4.

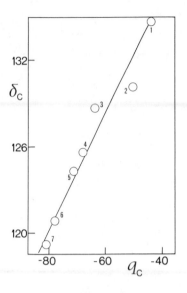

As one would anticipate from the similarity in the chemical nature of the compounds, the gross features revealed in Figures 4.1 and 4.2 are quite similar, namely as regards the increase in electron population at carbon resulting in an upfield shift. Not too much importance should be given to the difference between the slopes *a* calculated for the two series of compounds. Part of this difference is possibly due to the fact that the substituted benzenes were calculated using the "standard" STO–3G method, which is certainly a reasonable approach for this class of molecules, whereas the STO–3G remake of the Spiesecke–Schneider correlation has involved extensive geometry and scale factor optimizations, dictated by the diversity of the members of this series. In addition, one should consider that the Spiesecke–Schneider correlation involves cycles of different size, a circumstance which introduces an uncertainty regarding the validity (or lack of it) of interpreting chemical shift differences as a function of Mulliken charge density only, disregarding possible effects linked to shape. An indication of the overall influence of ring size on the quality of simple charge–shift correlations in this class of

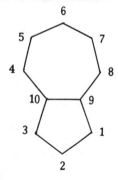

TABLE 4.5. Carbon Net Charges[18] and NMR Shifts of Azulene

Atom	q_π (me)	q_σ (me)	δ (ppm)[a]
1, 3	−102.2	−4.3	119.2
2	25.2	−77.2	137.7
4, 8	88.7	−117.6	136.9
5, 7	−40.7	−35.4	123.2
6	69.6	−101.5	137.4
9, 10	2.9	−3.5	140.8

[a] Converted data obtained from J.B. Stothers, "Carbon–13 NMR Spectroscopy", Academic Press, New York, NY, 1972.

TABLE 4.6. Atomic Populations (me) and Chemical Shifts (ppm from TMS) of Ethylenic Carbon Atoms

Molecule	Atom[a]	q_σ^{Mull}	q_π^{Mull}	$q_{\text{tot}}^{\text{Mull}}$	q_{tot} (Eq. 3.2)	δ^{b}
Ethylene	C (4)	−128.4	0.0	−128.4	−68.2	122.8
Propene	C–1 (2)	−120.4	−34.2	−154.6	−94.4	115.0
	C–2 (9)	−83.5	25.2	−58.3	−28.2	133.1
Isobutene	C–1 (1)	−112.2	−61.7	−173.9	−113.7	109.8
	C–2 (10)	−55.2	47.6	−7.6	−7.6	141.2
trans-Butene	C–2 (7)	−78.6	−7.7	−86.3	−56.2	125.8
cis-Butene	C–2 (6)	−77.6	−8.8	−86.4	−56.3	124.3
2-Methyl-2-butene	C–2 (8)	−49.8	12.3	−37.5	−37.5	131.4
	C–3 (3)	−73.8	−32.6	−106.4	−76.3	118.7
2,3-Dimethyl-2-butene	C–2 (5)	−50.2	−11.0	−61.2	−61.2	123.2

[a] The numbers in parentheses refer to the points in Figure 4.3. [b] Values of A.J. Jones and D.M. Grant, reported as personal communication in J.B. Stothers, "Carbon–13 NMR Spectroscopy", Academic Press, New York, NY, 1972. These results are in good agreement with those of G.B. Savitski, P.D. Ellis, K. Namikawa, and G.E. Maciel, J. Chem. Phys., **49**, 2395 (1968) (for propene); J.W. de Haan and L.J.M. van de Ven, Org. Magn. Res., **5**, 147 (1973) (for trans and cis-butene); R.A. Friedel and H.L. Retcofsky, J. Am. Chem. Soc., **85**, 1300 (1963) (for 2-methyl-2 butene).

compounds is offered by the study of azulene. From the results (Table 4.5) it appears that two distinct, parallel correlation lines are obtained, one for the 7-membered, the other for the 5-membered ring carbon atoms, indicating some sort of ring size- effect. When transposed on the scale of the correlation given in Figure 4.1, however, this effect is relatively modest. On these grounds we may regard that the shift–charge correlation presented in Figure 4.1 is, on the whole, reasonably good, mainly because it covers an important range of shift and charge results, but also that one should not attempt to extract more from it than it is capable of giving in terms of general trends. The results for substituted benzenes are more significant because they do not suffer from

FIGURE 4.3. Comparison of ^{13}C NMR shifts (ppm from TMS) with corrected carbon total $(\sigma + \pi)$ charges (in 10^{-3} electron units) of vinyl carbon atoms. The atom numbering is indicated in Table 4.6. The radius of the circles correspond to an uncertainty of 0.7 ppm or 3.5 me. (Reproduced from Ref. 27).

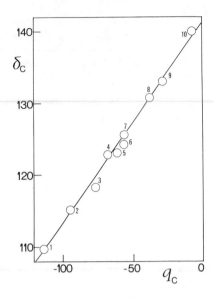

possible drawbacks linked to ring size and, indeed, their correlation is superior in quality to that given in Figure 4.1.

Finally, the STO–3G study[27] of simple ethylenic compounds had led to the results reported in Table 4.6. Contrasting, however, with the examples presented in Figures 4.1 and 4.2, we use here Eq. 3.2 for calculating the charges because the number of CH bonds formed by the sp^2 carbons is not the same in all cases. A multiple regression analysis indicated[27] $p = 31.8 \pm 4$ me, but a value of 30.12 is preferred (admittedly, in a somewhat arbitrary fashion) because this is the value found for sp^3 carbons (see Sect. 4). In any case, the ~ 30 me correction term is the only one permitting a monotonic dependence of ^{13}C NMR shifts on carbon charges. Incidentally, a detailed inspection[27] of these modified charges indicates that a methyl group is a better electron donor than hydrogen, in full agreement with the inductive effects discussed for the alkanes. Moreover, these results indicate a decrease in electron density at the point of substitution of methyl for hydrogen and an accumulation of negative charge at the β position. This polarization offers an explanation for Markownikoff's rule. The correlation between the sp^2 carbon chemical shifts and the modified atomic charges is presented in Figure 4.3, with $a = 291$ ppm/electron.

All the charge–shift relationships presented so far have one important point in common. They do, indeed, indicate a shift toward higher fields when the total $(\sigma + \pi)$ electron population at a carbon atom increases. Qualitatively, this is the trend which is usually invoked in shift vs. charge discussions although, in many cases, reference is made only to changes in π-electron densities. This type of trend is, however, no longer observed for the carbonyl-carbon atoms[28], as indicated by the results given in Table 4.7 and Figure 4.4.

TABLE 4.7. Mulliken Net Charges (me) and NMR Shifts (ppm from TMS) of Carbonyl-carbon Atoms

Molecule	q_σ	q_π	δ
CH_3CHO	100.3	111.3	199.6
C_2H_5CHO	93.1	111.1	202.4
$i\text{-}C_3H_7CHO$	86.8	111.1	204.3
1 $(CH_3)_2CO$	123.0	136.7	204.9
2 $CH_3COC_2H_5$	113.6	140.0	207.0
3 $CH_3CO\text{-}i\text{-}C_3H_7$	109.0	136.0	210.0
4 $(C_2H_5)_2CO$	104.2	143.2	209.4
5 $C_2H_5CO\text{-}i\text{-}C_3H_7$	99.6	139.2	212.3
6 $(i\text{-}C_3H_7)_2CO$	92.0	141.8	215.5

The NMR shifts are taken from C. Delseth and J.-P. Kintzinger, *Helv. Chim. Acta*, **59**, 466 (1976); *ibid.*, **59**, 1411 (1976). The charges are those given by "standard" STO–3G calculations[28].

For these atoms, any increase in total electron population is clearly reflected in a downfield shift, i.e., $a < 0$.

This change in the sign of a is, of course, most disturbing—a problem which shall be examined as follows. Traditionally, much of the discussion reported in the literature[1,2] about ^{13}C NMR shifts and electronic structure has related to aromatic systems, following Lauterbur's suggestion[29] that in these systems the shielding is governed primarily by the π-electron density at the carbon nuclei. Although the analysis presented here has emphasized relationships with total $(\sigma + \pi)$ atomic charges, there is no doubt that correlations with π-electron populations have their merits. For example, the

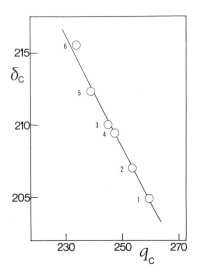

FIGURE 4.4. Comparison of ketone carbonyl–^{13}C NMR shifts (ppm from TMS) with net atomic charges, in 10^{-3} electron units. (From Ref. 28).

FIGURE 4.5. Carbon–13 NMR shifts (ppm from TMS) *vs.* π net charges (relative to 1 electron) for the aromatic compounds indicated in Table 4.3. The charges are in 10^{-3} electron units. (Reproduced from Ref. 18).

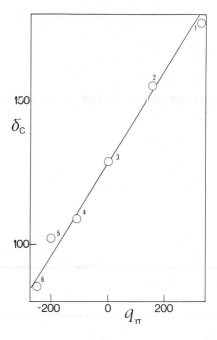

FIGURE 4.6. Correlation between the carbon–13 NMR shifts (ppm from TMS) of aromatic compounds (Table 4.3) and σ net charges (expressed in 10^{-3} electron units, relative to 5 electron).

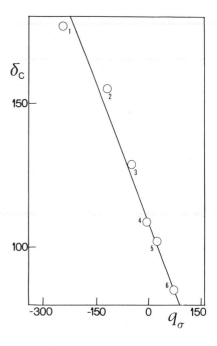

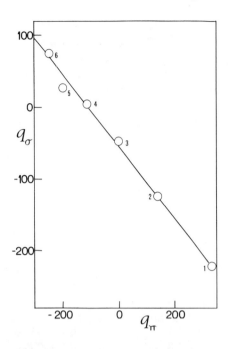

FIGURE 4.7. Comparison between σ and π net charges in aromatic compounds (Tables 4.2 and 4.3), expressed in 10^{-3} electron units. (Extracted from Ref. 18).

aromatics described in Table 4.3 yield the correlation with π charges presented in Figure 4.5 but it is also true that an equally good result is obtained if σ charges are used instead (Figure 4.6). This observation alone suffices to warn us that for aromatic (and, more generally, sp^2-carbon) systems the evaluation of the dependence of NMR shifts on electronic charge should not be restricted to π electrons only, with σ charges disregarded. In fact, the study of σ systems would otherwise come to an abrupt end before it has even started.

The reason why σ, π, and total ($\sigma + \pi$) charges yield correlations of similar quality for the aromatics is due to the linear decrease in σ population accompanying any increase in π electronic charge. Figure 4.7 illustrates this behavior for the hydrocarbons described in Table 4.2. Similarly, in the series of monosubstituted benzenes, the calculated changes in σ and π populations at the *para* carbon atom are accurately inversely proportional, as convincingly demonstrated in the Hehre–Taft–Topsom study[3]. The π population shows the greater change and the σ population seems to be consequently altered by $\sim 55\%$ in the opposite direction. Similar results are also obtained for the π and σ populations at the *meta* carbon atom, but these points show some scatter from linearity. However, most of the *meta* points fall close to the correlation line drawn for the *para* carbon atoms. Finally, a decrease in σ population accompanying a gain in π population is also observed for the vinyl carbon atoms indicated in Table 4.6. This result is best extracted from comparisons involving carbon atoms carrying the same number of hydrogen atoms (e.g., the CH_2 carbons of ethylene, propene and isobutene). In this manner we avoid the uncertainties associated with the Σp_{kl} terms (Eq. 1.5) and can reasonably hope that Mulliken charge differences reflect the trends in a correct fashion, at least qualitatively. Moreover, it must be borne in mind that with alkyl substitution the correct evaluation of π charges is biased by overlap with out-of-plane atoms, an effect which is difficult to account for in a reliable manner. The same argument applies to carbonyl-carbon atoms. However, while σ and π populations vary in opposite directions in the ketones, the π population of the aldehyde carbonyl-C atoms seems to remain fairly constant (Table 4.7).

Describing now, where appropriate, the observed changes in σ and π populations by Eq. 4.4

$$q_\sigma = mq_\pi + \text{const.} \tag{4.4}$$

it appears that Eq. 4.3 can be written as follows

$$\delta = a_\sigma q_\sigma + a_\pi q_\pi + \text{const.} \tag{4.5}$$

where

$$ma_\sigma + a_\pi = (m+1)a$$

represents the derivative $d\delta/dq_\pi$ (e.g., 160 ppm/e), which is now seen to account also for the fact that σ and π charges vary in opposite directions (from Eqs. 4.3, 4.4 and 4.5). Note that when Eq. 4.4 applies, the individual a_σ and a_π parameters cannot be obtained from simple regression analyses using Eq. 4.5 because q_σ and q_π are not independent variables. In σ systems ($q_\pi = 0$) or if Eq. 4.4 does not apply because $q_\pi = \text{constant}$, Eq. 4.3 takes the form

$$\delta = a_\sigma q_\sigma + \text{const.} \tag{4.6}$$

The shifts measured for saturated hydrocarbons, dialkylether-oxygen atoms (*vide infra*), and aldehyde carbonyl-carbon atoms indicate that $a_\sigma < 0$. Of course, there is no reason to assume that the values of a_σ and of a_π are the same in all systems. The observed charge *vs.* shift trends can be explained on a qualitative basis provided that

$$a_\sigma < 0, \quad a_\pi < 0, \quad |a_\sigma| > |a_\pi| \tag{4.7}$$

or $a_\sigma < 0$ and $a_\pi > 0$. The latter alternative presents no difficulty considering the inverse variations in σ and π populations (Tables 4.3–4.7) because both an increase in π population and the concurrent decrease in σ electronic charge would result in a shift toward higher fields. In this case, if the gain in π electrons is more important than the loss in σ population ($-1 < m < 0$ in Eq. 4.4), the upfield shift would be accompanied by a gain in total charge, i.e., $a > 0$ (ethylenic and aromatic hydrocarbons), but if $m < -1$, i.e., if the loss in σ electrons is more important than the gain in π population (carbonyl-carbon atoms) it would appear that $a < 0$ because of the actual decrease in total charge. On the other hand, it seems reasonable to assume that a_σ and a_π

have the same sign (as in Eq. 4.7), meaning that the same qualitative trend is expected from a variation in σ charge (at constant π population) or in π charge (at constant σ population). In that event, a downfield shift promoted by a gain in π electronic charge would be opposed by an inverse effect due to the concurrent loss in σ electrons. Provided that $|a_\sigma| > |a_\pi|$, an upfield shift can still result, even if the loss in σ population is less than the gain in π electrons (ethylenic and aromatic hydrocarbons). The upfield shift would, of course, be also observed anytime the loss in σ electrons exceeds the gain in π charge, for a net decrease in electron population, in which case $a < 0$.

Whichever, $a_\pi < 0$ or $a_\pi > 0$, turns out to be ultimately the "good" answer, this rationale is rooted in an anticipated difference in behavior between σ and π electrons. So, while it would have appeared that aromatic systems offer a simple access to the study of shift–charge relationships, particularly in situations where one does not have to worry about possible drawbacks arising from the use of Mulliken charges (see Eq. 1.5), it turns out that the way σ and π populations vary in opposite directions is of utmost importance—a circumstance which, if not properly recognized, is a source of difficulties concealed under a deceitful appearance of simplicity. The condition $a_\sigma < 0$ is, of course, dictated by the behavior observed for the aldehyde carbonyl-carbon atoms (Table 4.7). More significantly, however, the rationale offered for interpreting $a > 0$ and $a < 0$ situations is consistent with, and supported by, the results derived for a number of other systems, namely by sp^3-carbon and [17]O charge–shift relationships.

4. RELATIONSHIPS INVOLVING sp^3 CARBON ATOMS

The series of simple alkanes has been investigated at the extended Hückel level by Sichel and Whitehead[30] and at the STO–3G level[31] involving full optimization of all the geometry and exponent variational parameters. Qualitatively, the two methods lead to the same conclusions, which is a consequence of the similarities existing between charge results deduced from different methods (Chapters 2 and 3). Mulliken charge distributions are such, however, that a unique correlation with chemical shifts cannot be obtained for primary carbons and di-, tri- and tetraalkylsubstituted C atoms. Baird and Whitehead[32] suggested that part of the failure of these results may lie in an underestimation of the ionic characters, and charges therein, of the C—C and the C—H bonds. Here it is shown that the failure in obtaining a unique correlation between ^{13}C chemical shifts (for the alkanes) and carbon net charges can be attributed to an improper evaluation of the latter by using Mulliken's original definition. For the reasons given in detail in Chapter 3.1, full advantage is taken of Eqs. 1.3–1.5 using the simple approximation (Eq. 3.2):

$$q_C = q_C^{\text{Mulliken}} + N_{\text{CH}}p.$$

TABLE 4.8. Mulliken Orbital Populations of Selected Carbon Atoms from Optimized STO–3G Calculations (electron units)

Type of carbon	Molecule	Orbital population		
		$1s$	$2s$	$2p$
Primary	Ethane	1.9903	1.1459	2.8848
	Propane	1.9903	1.1491	2.8844
	Isobutane	1.9892	1.1513	2.8859
	Neopentane	1.9907	1.1540	2.8840
Secondary	Propane	1.9906	1.1522	2.8512
	Cyclohexane	1.9913	1.1550	2.8509
	Adamantane	1.9914	1.1559	2.8531
Tertiary	Isobutane	1.9916	1.1551	2.8199
	Adamantane	1.9917	1.1569	2.8193
Secondary	Cyclopentane	1.9918	1.1584	2.8618
	Cyclobutane	1.9907	1.1611	2.8726
	Cyclopropane	1.9910	1.1364	2.9603

The results used in the forthcoming discussion are based on the optimized *ab initio* Mulliken charge analyses[33-35] indicated in Table 4.8, which have led to the net charges investigated in Chapters 2 and 3 (namely, in Tables 2.8, 3.2 and 3.5).

A most instructive comparison involving Mulliken net charges is shown in Figure 4.8. The points for the CH_3-, CH_2-, and CH-carbon atoms, including those of cyclohexane and adamantane, lie on parallel, equidistant lines shifted from one another by ~ 30 me. On each regression line the carbon charge varies at the $2s$ level only, indicating that in this class of compounds an increase in $2s$-electron population promotes a ^{13}C downfield shift.

On the other hand, considering now the definition of charge which does

FIGURE 4.8. Comparison between the carbon–13 NMR shifts (ppm from TMS) of selected primary, secondary and tertiary sp^3 carbon atoms and Mulliken net atomic charges (in 10^{-3} electron units) derived from fully optimized STO–3G calculations (Tables 2.8 and 4.8, Refs. 33–35).

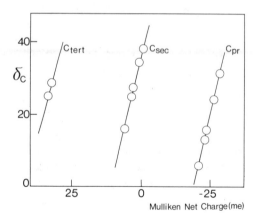

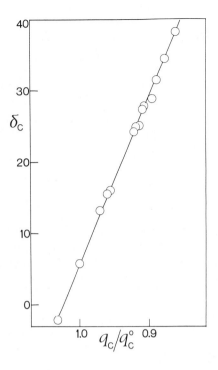

FIGURE 4.9. Correlation between ^{13}C NMR shifts (ppm from TMS) of sp^3 carbon atoms and modified atomic charges (Eq. 3.2) from fully optimized STO–3G calculations, with $p = 30.12$ me. The charge results are indicated in relative units, i.e., in terms of q_C/q_C°, and correspond to $n = -4.4122$ (see also Tables 3.5 and 4.9). Note that $q_C/q_C^\circ < 1$ indicates a decrease in positive charge, i.e., a gain in electron population.

not imply halving of overlap populations in heteronuclear cases (Eq. 3.2) we can write a linear relationship between chemical shifts and atomic charges as follows:

$$\delta = a(q_C^{\text{Mull}} + N_{\text{CH}}p) + b$$

and deduce p from a multiple regression analysis using Mulliken charges and experimental chemical shifts. The result, $p = 30.12$ me (from optimized STO–3G charges), can also be obtained from the "inductive formulas" (Table 2.4) and the calculation of the appropriate n value. The result, $n = -4.4122$, yields $p = 30.12$ me from Eqs. 3.3 and 3.4, and $q_C^\circ = $ modified net C charge in ethane $= 69.40$ me, at the optimized STO–3G level. These, of course, are the modified net atomic charges which were extensively discussed in Chapter 3. Expressing now the net charges on the scale of relative units, i.e., in terms of q_C/q_C° (see Table 3.5), the correlation with chemical shifts is as indicated in Figure 4.9. Note that this set of carbon charges, corresponding to an empirical partioning of CH overlap populations replacing their arbitrary halving, is the only one permitting a monotonic dependence of alkane–^{13}C NMR shifts on carbon charges. The situation is the same as that encountered with vinyl-carbon atoms (Table 4.6). Both for the alkane-sp^3 carbon atoms and the sp^2 carbons in alkyl-substituted ethylenes, one has to choose only between two alternatives, i.e., (i) virtually linear relationships with $p \simeq 30$ me or (ii) no relationship at all between ^{13}C shifts and theoretical atomic charges, with any $p \neq \sim 30$ me. Experimental partitionings of overlap populations at the optimized STO–3G level have also been made using ionization potentials[36] and energies of atomization[37] (Chapter 6.4), and carbon charges defined by Eq. 3.2. The n (and, therefore, p) values turn out to be the same in all cases. This point is particularly important in the forthcoming calculations of molecular energies (Chapter 6–9); since the same definition of charge satisfies the appropriate energy expressions on one hand, and ^{13}C NMR shifts on the other, we can use the latter for deriving the carbon charges required in energy calculations. The correlation shown in Figure 4.9 is described by Eq. 4.8[31]

$$\delta = -237.1 \left(\frac{q_C}{q_C^\circ} \right) + 242.64 \text{ ppm from TMS} \tag{4.8}$$

with a standard error of 0.3 ppm. The choice of tetramethylsilane as a standard of reference is, in itself, arbitrary and represents purely a matter of practical convenience. In fact, choosing ethane as the reference compound for defining alkane-carbon NMR shifts, Eq. 4.8 can be written as follows, with $\Delta q_C = q_C - q_C^\circ$,

$$\delta = -237.1 \left(\frac{\Delta q_L}{q_C^\circ} \right) \text{ ppm from ethane.} \tag{4.9}$$

The merit of Eq. 4.9 is to express clearly that the only empirical parameter required in the correlation between chemical shifts and sp^3-carbon net charges is the slope, $-237.1/q_C^\circ$ ppm/charge unit. This means that the ^{13}C NMR spectrum of adamantane, for example, can be deduced from the knowledge of only one alkane ^{13}C shift relative to ethane, e.g., from that of methane. Comparisons between observed and calculated results are presented in Table 4.9.

These equations also indicate that ^{13}C chemical shifts are extremely sensitive to small variations in charge. Indeed, a 1% change in q_C corresponds to varying δ by ~ 2.4 ppm. The standard deviation (0.3 ppm) is, therefore, small in comparison with the error in δ which would result from a 1% error in the estimate of q_C; the uncertainty in q_C can thus be estimated at $\sim 0.13\%$. As a consequence of the excellent agreement of Eq. 4.9 with experimental results, this equation can now be used to calculate carbon net charges when these are not known from previous work. Moreover, these charges (corresponding to $n = -4.4122$) can be used for deducing, by means of Eqs. 3.2– 3.4, the charges which would result from a STO–3G calculation involving full optimization of all (geometry and ζ scale factor) variational parameters. In conclusion, ^{13}C chemical shifts not only represent a severe test for the theory of charge distributions in alkanes but also represent a powerful means of obtaining C charges which would otherwise imply extremely lengthy optimized STO–3G calculations, often at or beyond the limits of their practical feasibility. These charges can then be used with confidence in studies of molecular properties involving charges, e.g., enthalpies of formation[38],

TABLE 4.9. Carbon Net Charges, Expressed as q_C/q_C(ethane), Calculated from the "Inductive Formulas" (Table 2.4, with $n = -4.4122$), STO–3G Theory and from Experimental ^{13}C NMR Shifts: Comparison between Experimental and Calculated Shifts

Molecule	Atom	q_C/q_C(ethane) *from*			δ, ppm from TMS	
		Ind. Form.	STO–3G	^{13}C *shift*	*Exptl.*	*Calcd.*
Methane	C	1.031	1.031	1.032	−2.1	−1.8
Ethane	C	1.000	1.000	0.999	5.8	5.5
Propane	C_{prim}	0.958	0.959	0.958	15.6	15.3
	C_{sec}	0.954	0.954	0.955	16.1	16.4
Butane	C_{prim}	0.968		0.968	13.2	13.1
	C_{sec}	0.920		0.918	25.0	24.5
Pentane	$C_{centr.}$	0.878		0.878	34.5	34.5
Isobutane	C_{prim}	0.922	0.922	0.921	24.3	24.0
	C_{tert}	0.915	0.915	0.917	25.2	25.7
Neopentane	C_{prim}	0.889	0.889	0.891	31.5	31.9
	C_{quat}	0.907	0.907	0.906	27.9	27.6
Cyclohexane	C		0.908	0.907	27.7	27.4
Adamantane	C_{sec}		0.862	0.862	38.24	38.3
	C_{tert}		0.896	0.902	28.75	30.2

The experimental shifts are reported in D.M. Grant and E.G. Paul, *J. Am. Chem. Soc.*, **86**, 2984 (1964) (methane-neopentane), D.K. Dalling and D.M. Grant, *J. Am. Chem. Soc.*, **89**, 6612 (1967), and G.E. Maciel, H.C. Dorn, R.L. Green, W.A. Kleschick, M.R. Peterson, and G.H. Wahl, *Org. Magn. Res.*, **6**, 178 (1974). The calculated shift values (Eq. 4.8) were deduced using STO–3G charges where available and the inductive charges for butane and pentane. The slight discrepancy observed for the tertiary carbons in adamantane is attributed to a lack of geometry optimization, dictated by reasons of computational feasibility[35].

energies of atomization[37], etc., which is to say that these properties can be deduced from ^{13}C NMR shifts.

The validity of Eq. 4.8, *viz.* Eq. 4.9, has been carefully established for linear and branched paraffins[31], cyclohexane and methylated cyclohexanes[34] and for molecules consisting of several cyclohexane rings in the chair conformation (namely, *trans*-decalin, *cis*-decalin, bicyclo[3.3.1]nonane, adamantane and methylated adamantanes) as well as in boat conformation (iceane and bicyclo[2.2.2]octane)[35]. No special effect seems to contribute to the chemical shift specifically because of the cyclic structure of cyclohexane. Similar conclusions are no longer true for smaller cycles (e.g., cyclopropane) which, not unexpectedly, grossly fail to obey Eq. 4.9. It is also true that the electronic structure of cyclopropane, for example, differs significantly from the pattern exhibited by the compounds satisfying Eq. 4.9. For the latter, the shift appears to be governed by Mulliken 2s populations (see Figure 4.8). The results of Table 4.8, on the other hand, indicate a regular increase of ~ 32.7 me in 2p population at carbon for each hydrogen atom attached to it, suggesting that $2p$ electrons largely make up for the p correction term for one CH bond (Eq. 3.2) so that, indeed, the change in total net charge appears to be mainly one of $2s$ electrons for the whole series of compounds (Figure 4.9). Cyclopropane, however, differs from the alkanes satisfying Eq. 4.9 in that its $2p$-electron population is ~ 110 me larger than that of other CH_2-carbon atoms, with an actual loss of $2s$ electrons[34]. Under these circumstances it is clear that cyclopropane cannot be considered on the same footing as the sp^3 hybridized carbon atoms obeying the charge–shift correlation expressed in Eq. 4.9, moreover as the validity of the approximation, Eq. 4.2, should also be questioned in this case, not to speak of the anticipated effects linked to the change in shape of the electron clouds. Similar conclusions apply also to cyclopentane and cyclobutane[18].

The merit of Eq. 4.9 lies not only in its accuracy but also in the fact that it is basis set independent, precisely because of the use of charges expressed in terms of $\Delta q_C / q_C^\circ$. Energy calculations (Chapter 6.4), on the other hand, indicate $q_C^\circ = 35.1$ me (see also Chapter 3.6). In this manner it follows from Eq. 4.9 that

$$\Delta q_C = -0.148 \delta_C (\text{me}) \qquad (4.10)$$

where $\Delta q_C = q_C - q_C^\circ$ is taken with reference to the ethane-carbon net charge and δ_C is the carbon shift relative to that of ethane (~ 5.8 ppm from TMS). This equation applies in charge calculations of acyclic and six-membered cyclic saturated hydrocarbons and of alkyl substituents in molecules containing functional groups. As demonstrated by Delseth and Kintzinger[39], correlations between the ^{13}C shifts of the carbon atoms in ROR′ ethers and those of the "parent" RCH_2R' hydrocarbons clearly reflect the close correspondence in the structure-related effects which govern electron distributions, indicating that the individual carbon atoms of the alkyl part of the ethers behave quite like hydrocarbon-C atoms. The same relation applies, at least as a valid approximation, also to the α-carbons in ethers, whose Mulliken $2p$ populations (from nonoptimized STO–3G calculations) increase by ~ 30 me for each added hydrogen atom, in line with the pattern exhibited by β-carbons and alkane-carbon atoms, showing that chemical shift differences appear to be primarily related to changes in $2s$ populations. Taking the α-carbon of diethylether as reference, Eq. 4.10 expresses Δq_C with respect to that atom and δ_C is the shift relative to the same atom (65.9 ppm from TMS[39]). Tentatively, Eq. 4.10 is also used for carbonyl α-carbon atoms (taking acetone as reference) although little can be said about the validity of this approach, except that energy results derived therefrom (Chapter 9.4) are satisfactory[40].

The general validity of Eq. 4.10 is of utmost importance in the description of the saturated hydrocarbon part of any organic molecule, keeping in mind, however, that it applies only to alkyl and cyclohexanic carbon atoms. A relation of similar accuracy is also known for the oxygen atoms in dialkylethers.

TABLE 4.10. Orbital Populations of the Oxygen Atoms in Dialkylethers

| Molecule | Orbital population (electron) | | | | |
	$1s$	$2s$	$2p_x$	$2p_y$	$2p_z$
$(CH_3)_2O$	1.99789	1.83322	1.95064	1.05848	1.45684
$CH_3OC_2H_5$	1.99790	1.83327	1.95069	1.06457	1.45960
$CH_3O\text{-}i\text{-}C_3H_7$	1.99790	1.83448	1.95088	1.06533	1.46467
$CH_3O\text{-}t\text{-}C_4H_9$	1.99789	1.83398	1.95065	1.06834	1.46575
$(C_2H_5)_2O$	1.99790	1.83322	1.95074	1.07078	1.46213
$C_2H_5O\text{-}i\text{-}C_3H_7$	1.99790	1.83449	1.95087	1.07161	1.46713
$C_2H_5O\text{-}t\text{-}C_4H_9$	1.99790	1.83401	1.95064	1.07465	1.46808
$(i\text{-}C_3H_7)_2O$	1.99790	1.83435	1.95012	1.07520	1.47250
$i\text{-}C_3H_7O\text{-}t\text{-}C_4H_9$	1.99790	1.83388	1.94995	1.07828	1.47336
$(t\text{-}C_4H_9)_2O$	1.99789	1.83387	1.94943	1.07564	1.47729

Results extracted from Ref. 28. The x axis is perpendicular to the (y, z) plane containing the COC atoms.

5. RELATIONSHIPS INVOLVING OXYGEN ATOMS

The Mulliken orbital populations deduced for dialkylether-oxygen atoms from standard STO–3G calculations indicate that the changes in population occur at the $2p$ level[28] (Table 4.10). The net atomic charges of the oxygen atoms illustrate the build-up of electronic charge with increasing electron-releasing ability of the substituents, reflecting the customary inductive effects. A comparison of the ionization potentials[41] with these charges yields a correlation, $IP = 0.0274q_O + 18.16$ eV (with an average error of 0.040 eV and a correlation coefficient of 0.9914), which reflects the expectation that electron withdrawal becomes easier as the oxygen atom becomes electron-richer. This result is worth mentioning because it illustrates a physical aspect linked to the ordering of the oxygen charges given by Mulliken's analysis, in full agreement with all known aspects related to the inductive effects of alkyl groups. In short, it can be assumed with confidence that the oxygen atom in, say, di-*tert*-butylether is electron-richer than that of dimethylether. This confidence is important in the following comparisons involving ^{17}O nuclear magnetic resonance shifts (Table 4.11, Figure 4.10).

These results indicate that in dialkylethers any gain in electronic charge at the oxygen atom is accompanied by a downfield ^{17}O NMR shift, i.e., $a_\sigma < 0$, which is also the trend exhibited by the sp^3 carbons discussed in Sect. 4.4. For future use in the calculation of the enthalpies of formation (and related quantities) of oxygen-containing compounds, we shall now examine more closely the results obtained for the ethers. In the first place, it is clear that the trends in electron populations at the ROR′ ether oxygen atoms reflect the electron-releasing (or withdrawing) abilities of the R, R′ alkyl

TABLE 4.11. ^{17}O NMR Shifts and Net Atomic Charges of Ether Oxygen Atoms from Standard STO–3G Calculations

Molecule	δ_O	q_O (me)
1 $(CH_3)_2O$	−52.5	−297.1
2 $CH_3OC_2H_5$	−22.5	−306.0
3 $CH_3O\text{-}i\text{-}C_3H_7$	−2.0	−313.3
4 $CH_3O\text{-}t\text{-}C_4H_9$	8.5	−316.6
5 $(C_2H_5)_2O$	6.5	−314.9
6 $C_2H_5O\text{-}i\text{-}C_3H_7$	28.0	−322.0
7 $C_2H_5O\text{-}t\text{-}C_4H_9$	40.5	−325.3
8 $(i\text{-}C_3H_7)_2O$	52.5	−330.1
9 $i\text{-}C_3H_7O\text{-}t\text{-}C_4H_9$	62.5	−333.4
10 $(t\text{-}C_4H_9)_2O$	76.0	−334.1

The chemical shifts (ppm from water) are extracted from Ref. 39. The numbering corresponds to the points in Fig. 4.10.

groups, just as is the case with the methylene carbons in RCH_2R' hydrocarbons. Since for the ether oxygen and the sp^3 carbon atoms it now appears that chemical shifts and atomic charges are linearly related to one another, it follows that the ^{17}O NMR shifts of the ROR′ ethers are expected to correlate with the methylene ^{13}C shifts of the corresponding RCH_2R' hydrocarbons. This is, indeed, the case, as demonstrated convincingly by Delseth and Kintzinger[39]. Now, for sp^3 carbon atoms, the equation relating their NMR shifts to their charge increments was derived on the basis of fully optimized STO–3G calculations. In order to deduce for ^{17}O nuclei an

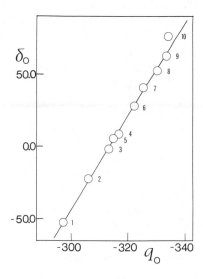

FIGURE 4.10. Correlation between ether ^{17}O NMR shifts (downfield from water) and Mulliken net atomic charges (in me units) from STO–3G calculations (extracted, in part, from Ref. 39). The numbering is indicated in Table 4.11.

expression which is on the same footing as Eq. 4.10, extensive geometry and scale factor optimizations were carried out for the dimethyl-, diethyl-, and diisopropyl ethers. Under these conditions, the oxygen STO–3G Mulliken charges of dimethylether (-267 me), diethylether (-295 me) and diisopropylether (-322 me) indicate that the δ_O vs. q_O slope is $\sim 1/1.8$ that of the corresponding δ_C vs. q_C slope, obtained from the same basis set, i.e.,

$$\frac{\Delta\delta_O}{\Delta q_O} \simeq \left(\frac{\Delta\delta_C}{\Delta q_C}\right)\left(\frac{1}{1.8}\right).$$

$$\Delta q_O = -0.267\delta_O \text{ (me)} \tag{4.11}$$

This relationship now enables a rapid estimate of ether-oxygen atomic charges on the scale which has proven satisfactory for accurately deriving charge-dependent molecular properties involving sp^3 carbon atoms.

As regards the carbonyl-oxygen atoms, however, the available information is not of the same quality as for the ether oxygens. An examination of results obtained for selected aldehydes and ketones[28] (Table 4.12) reveals that an electron enrichment at carbonyl oxygen atoms is accompanied by an upfield ^{17}O NMR shift, just as is the case for vinyl carbon atoms. However, because of the narrow range of Δq_O variations, lack of precision (~ 1–2 me) accompanying standard STO–3G calculations makes it difficult to assess how accurately carbonyl-^{17}O NMR shifts and charges are related to one another. As explained in Chapter 9.4, the approximation

$$\Delta q_O(\text{carbonyl}) \simeq 2.7\delta_O \tag{4.12}$$

appears to be justified on grounds of energy calculations but, of course, in this case our knowledge concerning Eq. 4.12 is limited by the accuracy of the experimental data used in this type of verification. On the other hand, a reliable theoretical assessment is presently out of reach because of the lack of information regarding heteronuclear π charges and their appropriate partitioning among the partners engaged in this type of bonding. The detailed charge analysis (Table 4.12) also reveals that, with increasing electron population, the gain in π-charge is more important than the loss in σ population. The observed upfield shift ($a < 0$) is thus explained in the same way as for the ethylenic and aromatic carbon atoms[18].

6. CHARGE ANALYSES

It is fair to conclude that the views expressed so far in this Chapter represent a significant step in the understanding of charge–shift relationships, although it is also clear that the number of new questions triggered in this manner surpasses those which have been dealt with. The information which has been collected here, while still modest in that it is presently limited to few classes of compounds, can be used in selected cases of particular interest in order to obtain possibly rough but realistic pictures of atomic charges. The examples

TABLE 4.12. Mulliken Charges and ^{17}O NMR Shifts of Carbonyl-oxygen Atoms (me, *viz*. ppm from water)

Molecule	q_σ	q_π	q_{tot}	δ
CH_3CHO	-103.7	-125.2	-228.9	592.0
C_2H_5CHO	-104.9	-124.7	-229.6	579.5
$i\text{-}C_3H_7CHO$	-105.4	-124.7	-230.1	574.5
$(CH_3)_2CO$	-104.5	-162.2	-266.7	569.0
$CH_3COC_2H_5$	-103.6	-165.9	-269.5	557.5
$CH_3CO\text{-}i\text{-}C_3H_7$	-106.3	-161.9	-268.1	557.0
$(C_2H_5)_2CO$	-102.7	-169.3	-272.1	547.0
$C_2H_5CO\text{-}i\text{-}C_3H_7$	-105.3	-165.4	-270.7	543.5
$(i\text{-}C_3H_7)_2CO$	-104.1	-169.6	-273.8	535

The NMR results are taken from Ref. 42.

discussed below yield results which prove useful in forthcoming calculations of molecular energies.

The calculation of the carbon and oxygen net charges is straightforward, using Eqs. 4.10–4.12 and the appropriate references q_C°(α-carbon), q_C°(ethane) = 35.1 me for carbons other than in α position, and q_O° for oxygen. The charges of C atoms other than adjacent to oxygen closely parallel those of the parent hydrocarbons[39,42] and it seems reasonable to anticipate a similar behavior also for the H atoms attached to them. In alkanes, the H and C charges are monotonically related to one another, with different lines for primary and secondary atoms. These lines are easily deduced using the "inductive" charges of Table 3.5 with q_C°(ethane) = 35.1 me, which ensures the correct scaling of atomic charges, giving (in me units) 35.1 (C), -11.70 (H) for ethane, 33.64 (C), -12.10 (H) for propane-CH_3, 32.37 (C), -12.47 (H) for isobutane-CH_3, and 31.20 (C), -13.05 (H) for neopentane-CH_3. For CH_2 charges, the values found are: 33.48 (C), -14.09 (H) for propane, 32.28 (C), -14.62 (H) for butane, and 30.82 (C), -15.01 (H) for the central CH_2 of *n*-pentane. In this manner it follows for di-*tert*-butylether ($\delta_{C\neq\alpha}$ = 26.0 ppm from ethane) that $q_{C\neq\alpha}$ = 31.25 and q_H = -13.02 me, on the average. Moreover, using $\delta_{C\alpha}$ = 7.7 ppm from the α-carbon of diethylether, $q_{C\alpha} = q_{C\alpha}^\circ - 1.14$ and (δ_O = 69.5 ppm from diethylether) $q_O = q_O^\circ - 18.56$ me. It follows from charge normalization that

$$q_O^\circ + 2q_{C\alpha}^\circ \simeq 67.7 \text{ me}$$

for the ether oxygen and α-carbon atoms. Similarly, one obtains for acetone ($\delta_{C\neq\alpha}$ = 22.2 ppm from ethane) that $q_{C\neq\alpha}$ = 31.81, q_H = -9.40 and

$$q_{C\alpha}^\circ + q_O^\circ \simeq -7.2 \text{ me}.$$

The problem is now one of evaluating the charges of the ether and acetone α-carbons, $q_{C\alpha}^\circ$. This is best done in terms of the difference

$$\Delta q_{C\alpha}^\circ = q_{C\alpha}^\circ - q_C^\circ(\text{ethane})$$

considering, however, that oxygen introduces an "extra" downfield shift at α-carbon (estimated[39] at ~ 41.7 ppm) which does not represent a carbon-charge effect. The latter is evaluated by subtracting tentatively 41.7 ppm from the observed α-carbon shifts (60, *viz.* 199 ppm from ethane for diethylether[39] and acetone[42], respectively) giving (Eq. 4.10) $\Delta q_{C_\alpha}^\circ \simeq -2.7$ me (ether) and ~ 23 me (acetone). A refinement based on energy calculations (Chapters 9.2–9.4) yields $\Delta q_{C_\alpha}^\circ \simeq -3.84$ me for diethylether and $\Delta q_{C_\alpha}^\circ \simeq -21.1$ me for acetone. Using these $\Delta q_{C_\alpha}^\circ$ values, it is found that $q_{C_\alpha}^\circ \simeq 31.26$ and $q_O^\circ \simeq 5.18$ me for diethylether, and $q_{C_\alpha}^\circ \simeq 14.0$ and $q_O^\circ \simeq -21.2$ me for acetone. The charges of H atoms attached to α-carbons are deduced from charge normalization.

This admittedly crude (but realistic) assignment of hydrogen charges is straightforward in most cases and a comparison of the trends in the sets of molecules studied in Chapter 9 facilitates their reasonable, balanced assessment. Note that this type of analysis is not required in actual energy calculations referring to chemical bonds which represent, by far, the leading energy terms. The individual hydrogen charges are, as a matter of fact, not needed in these calculations, but only their sum Σq_H, which is conveniently deduced from charge normalization (Chapter 9.2). The individual hydrogen net charges appear only in the evaluation of nonbonded interaction energies, whose contributions to atomization energies turn out to be very small indeed (Chapter 6.6). The important point concerning the estimates along the lines described above is that the uncertainties (usually < 0.1 me) have virtually no effect on the evaluation of nonbonded interactions.

7. CONCLUSIONS

In series of closely related compounds, it may occur that the change in total (dia- plus para-) magnetic shielding of atomic nuclei is nearly that of the *local* paramagnetic term, because of cancellation effects involving nonlocal dia- and paramagnetic contributions. This offers a justification for relationships between nuclear magnetic resonance shifts and local atomic populations, which are occasionally observed.

It is important, however, to consider the type ($2s$, $2p$, σ or π) of electrons which are responsible for the variations in atomic charges. Correlations between ^{13}C NMR shifts and atomic populations of aromatic compounds, for example, should not be interpreted in terms of π-electrons only, because the slope $d\delta/dq_\pi$ of shift *vs.* π-charge (i.e., the ~ 160 ppm/e value which is usually invoked) does not describe an intrinsic effect of π-charges on magnetic shielding but accounts also for the fact that σ and π charges vary in opposite directions in this class of compounds. The explicit consideration of the inverse variations of σ and π charges, where appropriate, offers an explanation for the observation that charge–shift correlations can have positive or negative slopes. It appears, indeed, that an increase in total electronic population is accompanied (*i*) by an upfield shift when the electron enrichment results from a gain in π-charge prevailing over the concurrent loss in σ-electrons (aromatic and vinyl-C, carbonyl-O atoms) or (*ii*) by a downfield shift when the increase in charge is dictated by that of the σ population (sp^3-C, carbonyl-C and dialkylether-O atoms).

It is important to be aware that both situations, i.e., up- or downfield shifts with increasing electron populations, are encountered, depending on the type of system under study. Charge–shift correlations established for a given series of molecules should not be indiscriminately assumed to be valid in all systems. In series of closely related compounds, however, charge–shift correlations can occasionally give valuable information about atomic charges, thus offering the possibility of studying large molecules which would otherwise lie outside the range of computational feasibility; whether honestly earned from lengthy (and expensive) theoretical calculations, or simply "borrowed" from critically established empirical correlations, a charge is a charge and a useful quantity as long as its reliability can be assessed. Its usefulness is best illustrated by energy calculations explicitly involving the charges of the bond-forming atoms.

In the future it appears desirable, therefore, to expand our knowledge of charge–NMR shift correlations to other classes of molecules, e.g., alcohols, nitrogen-containing compounds, small-ring hydrocarbons, etc., along the lines having led to the results described so far. In some cases, "standard" *ab initio* charge calculations may prove adequate (as for dialkylether oxygen atoms) because variations in charge in a given series of compounds may be so important that minor uncertainties due to the lack of extensive geometry and scale factor optimizations do not affect the final results to any significant extent. In other cases, however, most careful optimizations of all geometry and scale factors are a requirement (as for the sp^3 carbons in the alkane series) because the variations in charge are small and need to be known with precision. Finally, the role of the partitioning of overlap populations in heteronuclear situations must always be kept in mind—a point certainly requiring additional investigations which could be conducted along the general lines described for sp^3 and ethylenic sp^2 carbon atoms.

Current work[43] indicates that full optimizations are a must in the study of small cycles, as in tetracyclo[4.1.0.0^{2,4}.0^{3,5}]heptane. Indeed, nonoptimized

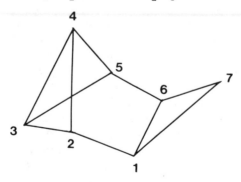

STO–3G calculations[44] suggest the 4-carbon to be electron richer than the carbon in 3 position, whereas the reverse is shown by optimized calculations[43] ($q_C^{Mulliken} = -150$ $vs. -138$ me, for C–3 and C–4, respectively). The latter result is in accord with the general trends indicating that a gain in local $2s$ charge is accompanied by a downfield ^{13}C NMR shift (δ 22.1 for C–3 and -4.0 ppm for C–4, from tetramethylsilane). These charge and shift results can be directly compared to one another because the local geometries (namely, the angles) about these two carbon atoms are very similar[43]. This, however, is not the case with other carbon atoms of 3-membered cycloalkanes, e.g., cyclopropane itself, a situation restricting the validity of simple charge–shift correlations in this class of compounds because of the changes in local geometry affecting the hybridization of carbon.

Del Re and coworkers[45] were concerned with the relation of s character in hybrids to bond angles and have considered hydridization as described by local orbitals, determined by requiring that hybrids on different atoms have minimal overlap unless they participate in the same bond. Alternate approaches are provided by the bond index of Wiberg[46] and by the Trindle-Sinanoğlu procedure[47] for the application of the physical criterion of Lennard-Jones and Pople[48,49], requiring that an electron in a localized orbital interacts maximally with the electron sharing that orbital. A good insight into the problems related to, and the possibilities offered by, the local orbital and bond index characterization of hybridization has been offered by Trindle and Sinanoğlu[50]: when a localized description of the wave function is possible, a situation which allows unambiguous definition of hybridization, the two methods give indistinguishable results. Calculated p characters are in good agreement with estimates (rooted in a work by Juan and Gutowsky[51]) derived from NMR coupling constants between carbon-13 nuclei and directly-bound protons[46,50]. Theoretical evaluations of hybridization, as well as estimates from experimental NMR coupling constants $J(^{13}CH)$, are anticipated to assist in future work on charge–shift correlations in cases suspected of presenting changes in local geometry capable of invalidating simple charge–shift relationships.

Charge–shift correlations like those described in this chapter are instrumental in the forthcoming highly accurate calculations of molecular binding energies. Before discussing this approach, however, it appears interesting to draw attention to another potential use of charge results—in estimates of correlation energies. Indeed, the energy of binding of atoms into a molecule is rigorously the sum of Hartree–Fock MO binding and electron correlation binding, the latter being a large fraction of the total binding energy. Now, the electron correlation term can be calculated by methods rooted in the theory of Sinanoğlu[52] of electron correlation in ground state molecules, in a simple semiempirical approximation[53] based on local atomic charges. A better understanding of the contribution of electron correlation to the binding energy is certainly of interest in theories of the near future, namely, in evaluations of "experimental" Hartree–Fock limits derived from experimental binding and calculated correlation energies. For molecules in their equilibrium geometry, on the other hand, the problem of molecular energies can be solved without an explicit separation into Hartree–Fock and correlation energies, in a simple electrostatic approach involving the charges of the bond-forming atoms, which can now be adequately retrieved from NMR spectral data. This topic occupies the next chapter.

REFERENCES

1. D.G. Farnum, *Adv. Phys. Org. Chem.*, V. Gold and D. Bethell, Eds., **11**, 123 (1975); G.L. Nelson and E.A. Williams, *Prog. Phys. Org. Chem.*, R.W. Taft, Ed., **12**, 229 (1976).
2. G.J. Martin, M.L. Martin, and S. Odiot, *Org. Magn. Res.*, **7**, 2 (1975).
3. W.J. Hehre, R.W. Taft, and R.D. Topsom, *Prog. Phys. Org. Chem.*, R.W. Taft, Ed., **12**, 159 (1976).
4. N.F. Ramsey, *Phys. Rev.*, **77**, 567 (1950); *ibid.*, **83**, 540 (1951); *ibid.*, **86**, 243 (1952). See also: W.H. Flygare, "Molecular Structure and Dynamics", Prentice-Hall Inc., Englewood Cliffs, NY, 1978.
5. R.M. Stevens, R.M. Pitzer, and W.N. Lipscomb, *J. Chem. Phys.*, **38**, 550 (1963).
6. W.N. Lipscomb, *Adv. Magn. Res.* J.S. Waugh, Ed., **2**, 137 (1966).
7. J.A. Pople, J.W. McIver, and N.S. Ostlund, *J. Chem. Phys.*, **49**, 2960 (1968).
8. J.A. Pople, *J. Chem. Phys.*, **37**, 53 (1962).
9. J. Tillieu, *Annales de Physique*, Vol. 2, **471**, 631 (1957).
10. J.R. Didry and J. Guy, *C.R. Acad. Sci.*, **239**, 1203 (1954).
11. W.N. Lipscomb, MTP *International Review of Science, Physical Chemistry*, Series One, Vol. 1., W. Byers Brown, Ed., Butterworths, London, 1972, p. 167.
12. R. Ditchfield, *Mol. Phys.*, **27**, 789 (1974).
13. K.A.K. Ebraheem and G.A. Webb, *Org. Magn. Res.*, **9**, 241 (1977); **10**, 70 (1977); G.I. Grigor and G.A. Webb, *ibid.*, **9**, 477 (1977); M. Jallali and G.A. Webb, *ibid.*, **11**, 34 (1978); K.A.K. Ebraheem, G.A. Webb, and M. Wittanwsky, *ibid.*, **11**, 27 (1978); K.A.K. Ebraheem and G.A. Webb, *J. Magn. Res.*, **25**, 399 (1977).
14. M.J. Stephen, *Proc. Roy. Soc.*, **A242**, 264 (1957).
15. R.M. Aminova, H.I. Zoroatskaya, and Yu. Yu. Samitov, *J. Magn. Res.*, **33**, 497 (1979).
16. E. Vauthier, S. Odiot, and F. Tonnard, *Can. J. Chem.*, **60**, 957 (1982).
17. J.A. Pople and D.L. Beveridge, "Approximate Molecular Orbital Theory", McGraw–Hill Book Company, New York, NY, 1970.
18. S. Fliszár, G. Cardinal, and M.-T. Béraldin, *J. Am. Chem. Soc.*, **104**, 5287 (1982).
19. A.A. Germer, *Theor. Chim. Acta*, **34**, 245 (1974).
20. M. Jallah–Heravi and G.A. Webb, *Org. Magn. Res.*, **13**, 116 (1980).
21. E. Vauthier, S. Fliszár, F. Tonnard, and S. Odiot, *Can. J. Chem.*, **61**, 1417 (1983).
22. H. Spiesecke and W.G. Schneider, *Tetrahedron Lett.*, 468 (1961).
23. G.A. Olah and G.D. Mateescu, *J. Am. Chem. Soc.*, **92**, 1430 (1970).
24. G.A. Olah, J.M. Bollinger, and A.M. White, *J. Am. Chem. Soc.*, **91**, 3667 (1969).
25. G.J. Ray, A.K. Colter, and R.J. Kurland, *Chem. Phys. Lett.*, **2**, 324 (1968).
26. E.A. LaLancette and R.E. Benson, *J. Am. Chem. Soc.*, **87**, 1941 (1965).
27. H. Henry and S. Fliszár, *J. Am. Chem. Soc.*, **100**, 3312 (1978).

28. M.-T. Béraldin, E. Vauthier, and S. Fliszár, *Can. J. Chem.*, **60**, 106 (1982).

29. P.C. Lauterbur, *J. Am. Chem. Soc.*, **83**, 1838 (1961).

30. J.M. Sichel and M.A. Whitehead, *Theor. Chim. Acta*, **5**, 35 (1966).

31. S. Fliszár, A. Goursot, and H. Dugas, *J. Am. Chem. Soc.*, **96**, 4358 (1974).

32. N.C. Baird and M.A. Whitehead, *Theor. Chim. Acta*, **6**, 167 (1966).

33. G. Kean and S. Fliszár, *Can. J. Chem.*, **52**, 2772 (1974).

34. R. Roberge and S. Fliszár, *Can. J. Chem.*, **53**, 2400 (1975).

35. G. Kean, D. Gravel, and S. Fliszár, *J. Am. Chem. Soc.*, **98**, 4749 (1976).

36. H. Henry and S. Fliszár, *Can. J. Chem.*, **52**, 3799 (1974).

37. S. Fliszár, *J. Am. Chem. Soc.*, **102**, 6946 (1980).

38. H. Henry, S. Fliszár, and A. Julg, *Can. J. Chem.*, **54**, 2085 (1976); S. Fliszár and J.-L. Cantara, *Can. J. Chem.*, **59**, 1381 (1981).

39. C. Delseth and J.-P. Kintzinger, *Helv. Chim. Acta*, **61**, 1327 (1978).

40. S. Fliszár and M.-T. Béraldin, *Can. J. Chem.*, **60**, 792 (1982).

41. K. Watanabe, T. Nakayama, and J. Mottl, *J. Quant. Spectrosc. Radiat. Transfer*, **2**, 369 (1962).

42. C. Delseth and J.-P. Kintzinger, *Helv. Chim. Acta*, **59**, 466 (1976); **59**, 1411 (1976).

43. M.-T. Béraldin, J. Bridet, and S. Fliszár, unpublished results.

44. M. Christl, *Chem. Ber.*, **108**, 2781 (1975).

45. A. Veillard and G. Del Re, *Theor. Chim. Acta*, **2**, 55 (1964); G. Del Re, U. Esposito, and M. Carpentieri, *Theor. Chim. Acta*, **6**, 36 (1966).

46. K.B. Wiberg, *Tetrahedron*, **24**, 1083 (1968).

47. C. Trindle and O. Sinanoğlu, *J. Chem. Phys.*, **49**, 65 (1968).

48. J.E. Lennard-Jones and J.A. Pople, *Proc. Roy. Soc. (London)*, **A220**, 446 (1950); *ibid.*, **A210**, 190 (1951).

49. C. Edmiston and K. Ruedenberg, *Rev. Mod. Phys.*, **35**, 457 (1963); *J. Chem. Phys.*, **43**, S97 (1965).

50. C. Trindle and O. Sinanoğlu, *J. Am. Chem. Soc.*, **91**, 853 (1969).

51. C. Juan and H.S. Gutowsky, *J. Chem. Phys.*, **37**, 2198 (1962).

52. O. Sinanoğlu, *J. Chem. Phys.*, **33**, 1212 (1960); *ibid.*, **36**, 706 (1962); *Advan. Chem. Phys.*, **6**, 315 (1964).

53. O. Sinanoğlu and H.Ö. Pamuk, *J. Am. Chem. Soc.*, **95**, 5435 (1973), and references cited therein.

CHAPTER 5

The Molecular Energy, A Theory of Electron Density

1. INTRODUCTION

The simplest "theoretical" description of a molecule, its structural formula featuring the chemical bonds, still represents the single most valuable material support for the discussion of a number of problems regarding chemical transformations and molecular properties. Because of the undisputed qualities of this type of description, much can be gained (namely, in simplicity) with the use of a theoretical approach retracing the essential features linked to the concept of chemical bonds.

While, of course, a molecule is a distinct entity which can be characterized in a number of manners, the constituent atoms retain much of their atomic properties. In fact, a molecule can be considered as a collection of "atoms" with energies differing from their free state values. These energy differences are, for each individual atom, a measure for the process of becoming part of the molecule and contain a portion of the molecular binding energy. Alternatively, in an entirely equivalent description, molecules can be regarded as assemblies of chemical bonds whose energies depend in a subtle way on the bond-forming atoms, namely, on their electron populations. This type of description, resulting from an appropriate energy partitioning, leads to a bond-by-bond insight into the various factors contributing to the stability of a molecular edifice and appears to be most suited for its analysis in terms of electron distributions.

Following a brief thermodynamic analysis showing that the energy of atomization of a molecule in its hypothetical vibrationless state at 0 K is the quantity of choice for studying molecular energies, and describing how this quantity can be derived from experimental data for use in comparisons with theory, we examine the relationships between energy components in isolated atoms and in molecules. In this manner, it becomes possible to define the energy of an atom in a molecule and that of the chemical bond in terms of basic nuclear–electronic and nuclear–nuclear interaction energies, thus reducing the problem to its simplest electrostatic expression. Finally, the role of the electronic charges carried by the bond-forming atoms is described in detail because these are, ultimately, responsible for virtually the entire energy variation accompanying isodesmic structural changes, thus permitting conformational analyses in terms of electron distributions.

2. THE MOLECULE IN ITS HYPOTHETICAL VIBRATIONLESS STATE

The atomization

$$\text{Molecule} \rightarrow n_1 A_1 + n_2 A_2 + \cdots$$

of a molecule into its constituent n_1 atoms A_1, n_2 atoms A_2, ... provides a direct insight into fundamental aspects of molecular energies. The relevant thermodynamic information is usually expressed in terms of enthalpy of formation, ΔH_f, or enthalpy of atomization, ΔH_a, of the molecule under consideration, i.e.,

$$\Delta H_a = \sum n_i \Delta H_f(A_i) - \Delta H_f \tag{5.1}$$

where the $\Delta H_f(A_i)$'s are the enthalpies of formation of the gaseous atoms A_i. A number of schemes viewing the molecule as an assembly of chemical bonds or groups and the "molecular energy" as a sum of appropriately defined contributions has been proposed for calculating ΔH_a[1-4]. Essentially empirical in nature, these additivity schemes involve in all cases a heavy parametrization which does, on one hand, ensure the success of the method but, on the other, obscures the fundamental physical reasons which are at the origin of molecular stabilities; once the principle of describing a particular topological situation by a parameter is accepted, the success of this type of approach depends on the number of different structural features which are considered and parametrized in order to cover an adequate body of compounds. For example, among the most successful methods, that proposed by Allen et al.[3] demonstrated on empirical grounds a requirement for seven parameters; then the heat of formation of all saturated acyclic hydrocarbons with eight carbon atoms or less are correlated with an average deviation of ~ 0.30 kcal/mol.

In the study of isolated molecules, we refer more conveniently[5] to the *energy* of atomization ΔE_a, i.e.,

$$\Delta E_a = \Delta H_a - \left(\sum n_i - 1 \right) RT. \tag{5.2}$$

Both ΔE_a and ΔH_a are considered at some temperature, usually 25°C, i.e. under working conditions which are of real practical interest. However, when used in the study of molecular properties, these quantities contain the seed for unnecessary difficulties, such as those arising from internal rotations which are more or less free in some cases and hindered in others. This is simply avoided by studying the molecules at 0 K. Moreover, zero-point vibrational energies are to be taken explicitly into account, as these energies, like the thermal ones, cannot fairly be apportioned among bonds or atoms in a molecule since they are not truly additive properties nor can they be regarded as a part of chemical binding. These reasons lead us to study a molecule in its hypothetical vibrationless state at 0 K, whose atomization energy is ΔE_a^*.

This energy measures the difference, $\Delta E_a^* = \Sigma_i E_i^{at} - E_{mol}$, between the total energy of the isolated atoms and that of the motionless ground-state molecule. It represents the quantity of prime interest in the study of molecules at equilibrium because it can be expressed in terms of fundamental contributions, i.e., nuclear–electronic, nuclear–nuclear and electron–electron interaction energies. The relation between ΔE_a^* and ΔE_a is described by Eq. 5.3

$$\Delta E_a = \Delta E_a^* - \sum F(v_i, T) + \frac{3}{2}\left(\sum n_i - 2\right) RT \qquad (5.3)$$

which states that the energy of atomization at, say, 25°C is that of the hypothetical vibrationless molecule at 0 K *less* the sum $\Sigma F(v_i, T)$ (over $3\Sigma n_i - 6$ degrees of freedom) of vibrational energy corresponding to the fundamental frequencies v_i, which is already present in the molecule at 25°C. The term $3(\Sigma n_i - 2)RT/2$ accounts for the formation of Σn_i atoms with translational energy and the disappearance of one nonlinear molecule with 3 translational and 3 rotational $RT/2$ contributions.

It follows from Eqs. 5.1–5.3 that

$$\Delta E_a^* = \sum_i n_i \left[\Delta H_f^\circ(A_i) - \frac{5}{2}RT\right] + \sum F(v_i, T) + 4RT - \Delta H_f^\circ \qquad (5.4)$$

where all enthalpies are now referred to standard conditions (gas, 298.15 K). The vibrational energy may be separated into a zero-point energy term (ZPE) and a thermal vibrational energy term (E_{vibr}). It is, therefore,

$$\sum F(v_i, T) + 4RT = ZPE + E_{vibr} + 4RT.$$

On the other hand, $E_{vibr} + 4RT$ is the increase in enthalpy ($H_T - H_0$) of nonlinear molecules due to their warming up from $T = 0$ to $T = T$. Consequently, Eq. 5.4 can now be written as follows

$$\Delta E_a^* = \sum_i n_i \left[\Delta H_f^\circ(A_i) - \frac{5}{2}RT\right] + ZPE + (H_T - H_0) - \Delta H_f^\circ \qquad (5.5)$$

This analysis of the energy components illustrates that there is long way to go from the knowledge of the behavior of hypothetical motionless mole-

TABLE 5.1. Energy Components (Eq. 5.5) of Selected Molecules (kcal/mol)

Molecule	ΔH_f° (gas, 298.15)	ZPE	H_T-H_0	ΔE_a^*
Ethane	−20.04	45.16	2.86	710.54
Pentane	−35.00	97.20	5.63	1592.20
Neopentane	−40.27	96.27	5.03	1595.94
Hexane	−39.96	114.37	6.62	1885.95
2,3-Dimethylbutane	−42.49	113.73	5.92	1887.14
Cyclohexane	−29.50	103.30	4.24	1760.82
Adamantane	−30.65	148.49	5.05	2688.05
Ethanal	−39.73	33.58	3.07	675.69
Propanone	−51.90	50.59	4.00	976.43
Dimethylether	−43.99	49.91	2.74	797.07
Diethylether	−60.26	82.82	4.94	1389.87
Tetrahydropyran	−53.39	88.83	4.04	1557.45

Details and additional examples are reported in: H. Henry, G. Kean, and S. Fliszár, *J. Am. Chem. Soc.*, **99**, 5889 (1977) and S. Fliszár and M.-T. Béraldin, *Can. J. Chem.*, **60**, 792 (1982). The appropriate $\Delta H_f^\circ(A_i)$ values are taken from D.R. Stull and G.C. Sinke, *Adv. Chem. Ser.*, **18** (1956).

cules to the understanding of their actual behavior at some temperature T. Postponing the questions about the derivation of accurate ΔE_a^*'s, it appears that a real problem is one of estimating the vibrational contributions in a simple manner because the appropriate spectroscopic data are, unfortunately, relatively scarce. Zero-point energies are obtained from vibrational spectra using experimental frequencies whenever available, while the non-observed frequencies are extracted from data calculated by means of an appropriate force-field model. In the harmonic oscillator approximation, the zero-point energy is

$$\text{ZPE} = N_{\text{Av}} \sum_i \frac{1}{2} h v_i \tag{5.6}$$

and the thermal vibrational energy is obtained from Einstein's equation

$$E_{\text{vibr}} = N_{\text{Av}} \sum_i h v_i \left(\frac{\exp(-h v_i/kT)}{1 - \exp(-h v_i/kT)} \right) \tag{5.7}$$

from which $H_T - H_0 = E_{\text{vibr}} + 4RT$ is readily deduced. Equations 5.5–5.7 are instrumental in the calculation of experimental ΔE_a^* atomization energies, for use in comparisons with their theoretical counterparts. Selected examples are presented in Table 5.1, which illustrate the magnitudes of the different energy components discussed above. They were calculated using standard gas enthalpy values at 298.15 K and $\Delta H_f^\circ(C) = 170.89$, $\Delta H_f^\circ(H) = 52.09$, and $\Delta H_f^\circ(O) = 59.54$ kcal/mol.

What these examples really indicate is that the study of molecules in their hypothetical vibrationless state ultimately leads to conventional enthalpy results (say, at 25°C) provided, of course, that adequate information about the vibrational parts can be collected. In fact, reversing the problem, accurate ΔE_a^*'s derived from theory and the appropriate enthalpies of formation give access to $ZPE + H_T - H_0$ data without having recourse to spectroscopic analyses, and are at the origin of a rich generation of molecular vibrational energies[6]. The ΔE_a^* energy is clearly the quantity of choice for isolating in a clear perspective the fundamental factors responsible for chemical binding, moreover, as it can be exactly expressed in terms of, and thus benefit from, quantum mechanical information. From thereon, however, a major part of the theory describing ΔE_a^* is developed in the spirit of the Hellmann–Feynman theorem, which shows that a consideration of classical electrostatic interactions suffices to determine the energy of a molecular system without the need for *explicit* inclusion of quantum mechanical contributions[7]. In a way, we shall take full advantage of quantum chemistry without really using it, at least not in the most customary fashion.

The relationships between energy components in isolated atoms and in molecules are now examined since both atomic and molecular energies are involved in the calculation of ΔE_a^*.

3. ENERGY COMPONENTS IN ISOLATED ATOMS

The potential energy V is made up from nuclear–electronic (V_{ne}) attraction energies and electron–electronic (V_{ee}) repulsions and the total atomic energy E includes, of course, also the electronic kinetic energy T, giving $E = T + V$. By applying the virial theorem, E is conveniently expressed in terms of potential energies only, i.e., $E = (V_{ne} + V_{ee})/2$. A considerable further simplification is achieved with the use of the ratio K_k^{at} defined by the equation

$$E_k(\text{free atom}) = K_k^{at} V_{ne}(\text{free atom}) \qquad (5.8)$$

relating the total energy E_k of a free atom k to its nuclear–electronic potential energy. In this electrostatic description, K_k^{at} is necessarily a constant for each individual atom and this is, ultimately, all that is required for treating the "atomic part" in the forthcoming description of ΔE_a^*. For hydrogen, K_H^{at} is obviously 1/2 because of the virial theorem and because $V_{ee} = 0$. The noteworthy point is that for all atoms other than hydrogen K_k^{at} always approaches the value 3/7. This remarkable result, illustrated by selected examples in Table 5.2, has been extensively described by Fraga[8] on an empirical basis and has triggered a number of theoretical investigations[9–14], namely, studies by Parr, Politzer et al., which show that the energy expression corresponding to the Thomas–Fermi model for an atom yields $E = (3/7)V_{ne}$. Incidentally, the K_k^{at} parameter of Eq. 5.8 can also be deduced from atomic total and Hartree–Fock orbital energies, by means of an equation derived in the next section which applies to both molecules and isolated atoms.

The V_{ne} potential energy is, of course, governed by the electron density

TABLE 5.2. Relationship between Total and Nuclear–Electronic Energies of Isolated Atoms (Eq. 5.8), in a.u.

Atom k	E_k (free atom)	V_{ne} (free atom)	V_{ne}/E_k
He	−2.8617	−6.7491	2.3584
Li	−7.4328	−17.1465	2.3069
Be	−14.5731	−33.6353	2.3080
B	−24.5291	−56.8973	2.3196
C	−37.6888	−88.1372	2.3386
N	−54.4011	−128.3518	2.3594
O	−74.8096	−178.0752	2.3804
F	−99.4096	−238.6683	2.4009
Ne	−128.5474	−311.1340	2.4204
Na	−161.8599	−389.7368	2.4079
Mg	−199.6154	−479.0470	2.3998
Al	−241.8775	−578.5018	2.3917
Si	−288.8550	−689.4163	2.3867
P	−340.7192	−812.2229	2.3838
S	−397.5061	−946.8970	2.3821
Cl	−459.4829	−1094.3544	2.3817
A	−526.8189	−1255.0608	2.3823

The atomic wave functions are extracted from: C. Froese Fischer, *At. Data Nucl. Data Tables* **4**, 301 (1972); *ibid.*, **12**, 87 (1973). Additional results are reported in: P. Politzer, *J. Chem. Phys.*, **70**, 1067 (1979).

$\rho(r)$, which is a function of r, the distance from the nucleus with charge Z. Maximum at the nucleus, $\rho(r)$ decreases monotonically and is well represented as a continuous piecewise exponential function of r with as many different exponential regions as there are principal quantum numbers[15]. The results for neon, depicting Hartree–Fock electron densities, are a typical example (Fig. 5.1a). The number of electrons at any given distance from the nucleus is, assuming spherical symmetry, $\rho(r) \cdot 4\pi r^2 dr$. The radial distribution function $R(r) = 4\pi r^2 \rho(r)$ for neon is depicted in Fig. 5.1b. Regarding the description of $\rho(r)$ vs. r, in an actual or Hartree–Fock atom the transitions from one exponential to the next occur over certain intervals and cannot be associated with single points. Contrasting with this situation, the minima in the radial density $R(r)$ are well-defined and physically meaningful points, which allow one to separate an atom into different exponential regions[9,15].

On these grounds, Politzer and Parr[9] have proposed that the minimum in the radial density function at the point r_m defines a boundary surface separating core and valence regions of first-row atoms. With this definition of the valence region, a valid estimate of its electronic energy can, indeed, be obtained with the formula[10,16-18]

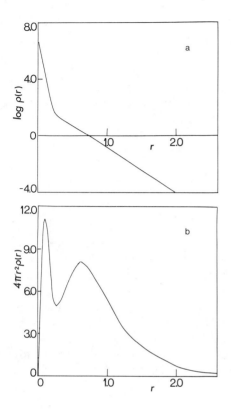

FIGURE 5.1. a) Electron density of neon. b) Radial distribution for neon. (Extracted from Ref. 15).

$$E_N = -\left(\frac{12\pi}{7}\right)(Z - N_1) \int_{r_m}^{\infty} \rho(r)r\,\mathrm{d}r$$

in which N_1 ($\simeq 2$ electrons) is the number of electrons in the core region of an atom with nuclear charge Z and $K_k^{at} = 3/7$. The core-valence separation in larger atoms[19] has been treated in a similar way, where r_m is then the outermost minimum in the radial density function. The success of this type of approximation justifies Parr's statement that "we may expect the atomic cores to be basic, transferable entities in the simple theories of the future". As explained further below, they do, indeed, play a role in the calculation of molecular energies.

4. RELATIONSHIPS BETWEEN ELECTRONIC, ORBITAL, AND TOTAL ENERGIES IN MOLECULES

The nucleus of an atom k being part of a molecule experiences a nuclear–electronic potential attraction energy which involves all the electrons of the molecule, i.e.,

$$V_{\mathrm{ne}}(k, \mathrm{mol}) = -Z_k \int \frac{\rho(\mathbf{r})}{|\mathbf{r} - \mathbf{r}_k|} \mathrm{d}\mathbf{r}.$$

Moreover, this nucleus is also subject to the nuclear–nuclear repulsion energy $Z_k \Sigma_{l \neq k} Z_l / r_{lk}$ due to all the other nuclei which are present in the molecule, with $r_{lk} =$ distance between nuclei k and l. The total (nuclear–electronic and nuclear–nuclear) potential energy involving Z_k is thus

$$V(k, \mathrm{mol}) = V_{\mathrm{ne}}(k, \mathrm{mol}) + Z_k \sum_{l \neq k} \frac{Z_l}{r_{lk}} \tag{5.9}$$

where the sum over l runs over all nuclei but k. Note that the sum over all atoms k of the molecule is

$$\sum_k V(k, \mathrm{mol}) = V_{\mathrm{ne}} + 2V_{\mathrm{nn}} \tag{5.10}$$

where V_{ne} and V_{nn} are, respectively, the total molecular nuclear–electronic and nuclear–nuclear potential energies. The total energy of the molecule is, of course, $E_{\mathrm{mol}} = T + V_{\mathrm{ne}} + V_{\mathrm{ee}} + V_{\mathrm{nn}}$, with $T =$ kinetic energy of the electrons and $V_{\mathrm{ee}} =$ electron–electron repulsion energy.

So far we have enumerated all the components from which the total molecular energy is made up. The kinetic energy is the first one to go in the following treatment because we consider only molecules in their equilibrium geometry, satisfying the virial theorem $E_{\mathrm{mol}} = -T$. It is the potential energy $V(k, \mathrm{mol})$ which turns out to be the central quantity in the description of "local" components resulting from a partitioning of the total molecular energy. Indeed, consider

$$E_{\mathrm{mol}} = E = \langle \psi_{\mathrm{mol}} | H_{\mathrm{mol}} | \psi_{\mathrm{mol}} \rangle$$

where H_{mol} and ψ_{mol} are, respectively, the appropriate Hamiltonian and the molecular ground-state wave function. The derivative $\partial E / \partial Z_k$ with respect to the nuclear charge Z_k of the kth atom in the molecule may be calculated leaving the internuclear distances and the number of electrons unchanged. It follows from the Hellmann–Feynman theorem that

$$\frac{\partial E}{\partial Z_k} = \left\langle \psi_{\mathrm{mol}} \left| \frac{\partial H_{\mathrm{mol}}}{\partial Z_k} \right| \psi_{\mathrm{mol}} \right\rangle$$

$$= \left\langle \psi_{\mathrm{mol}} \left| -\sum_i r_{ik}^{-1} + \sum_{l \neq k} Z_l / r_{lk} \right| \psi_{\mathrm{mol}} \right\rangle.$$

The sum over i runs over all electrons in the molecule and the corresponding integral is $V_{\mathrm{ne}}(k, \mathrm{mol}) / Z_k$. Using Eqs. 5.9 and 5.10 it follows that

$$Z_k \frac{\partial E}{\partial Z_k} = V(k, \mathrm{mol}) \tag{5.11}$$

and

$$\sum_k Z_k \left(\frac{\partial E}{\partial Z_k} \right) = V_{\mathrm{ne}} + 2V_{\mathrm{nn}}. \tag{5.12}$$

On the other hand, the total energy E of a molecule in its equilibrium geometry can be expressed by applying the virial theorem, i.e.,

$$2E = V_{ne} + V_{ee} + V_{nn}. \tag{5.13}$$

Consequently, one obtains from Eqs. 5.12 and 5.13 that

$$2E - \sum_k Z_k \left(\frac{\partial E}{\partial Z_k} \right) = V_{ee} - V_{nn}. \tag{5.14}$$

The equations derived so far apply equally well to exact as to Hartree–Fock wave functions which, of course, satisfy both the virial and the Hellmann–Feynman theorems. Let us now consider Hartree–Fock orbital energies ε_i^{orb} with occupations v_i. The interelectronic V_{ee} repulsions are counted twice in the computation of the sum $\sum_i v_i \varepsilon_i^{orb}$ of occupied orbital energies, whereas the nuclear–nuclear repulsions are obviously not included, a fact which is accounted for by the Hartree–Fock formula

$$E_{HF} = \sum_i v_i \varepsilon_i^{orb} - (V_{ee} - V_{nn})$$

which becomes, introducing Eq. 5.14,

$$3E_{HF} - \sum_k Z_k \frac{\partial E_{HF}}{\partial Z_k} = \sum_i v_i \varepsilon_i^{orb}. \tag{5.15}$$

Equation 5.15 is an identity in Hartree–Fock theory provided, of course, that all quantities are evaluated at equilibrium geometry. At this stage we define a new quantity

$$\gamma = \frac{1}{E_{HF}} \sum_k Z_k \left(\frac{\partial E_{HF}}{\partial Z_k} \right) \tag{5.16}$$

and rewrite Eq. 5.15 as follows[20,21]

$$(3 - \gamma) E_{HF} = \sum_i v_i \varepsilon_i^{orb}. \tag{5.17}$$

Finally, it is readily deduced from Eqs. 5.12 and 5.16 that[21]

$$E_{HF} = \frac{1}{\gamma} (V_{ne} + 2V_{nn}). \tag{5.18}$$

It is clear that Eqs. 5.17 and 5.18 are exact in Hartree–Fock theory for molecules at equilibrium. They are useful working formulas in the forthcoming developments. Inspection of Eq. 5.18 reveals that the explicit V_{ee} repulsion term has now also disappeared from the energy expression. The beauty of it is only obscured by the apparently somewhat mysterious new parameter γ, but this situation is just a temporary one. Postponing momentarily the detailed physical description of γ, a good insight can be gained from the numerical analysis presented in Table 5.3 and the following rough first approximations suggested by it.

Hartree–Fock calculations made for a great variety of molecules show

TABLE 5.3. Energy Components (a.u.) and γ Values of Selected Molecules from GTO $(9s5p|6s) \rightarrow [5s3p|3s]$ Calculations

Molecule	E_{HF}	V_{ne}	V_{nn}	$\Sigma_i \nu_i \varepsilon_i^{orb}$	γ	
					Eq. 5.17	Eq. 5.18
CH_4	-40.1873	-120.0380	13.5245	-27.6090	2.3130	2.3139
C_2H_6	-79.2096	-268.1916	42.2663	-53.9792	2.3185	2.3186
C_3H_8	-118.2337	-439.4772	82.5597	-80.3580	2.3203	2.3205
$i\text{-}C_4H_{10}$	-157.2587	-634.2056	134.5693	-106.7563	2.3211	2.3214
$c\text{-}C_3H_6$	-117.0236	-423.9282	76.0498	-79.3177	2.3222	2.3229
C_2H_4	-78.0166	-248.4619	33.6233	-52.8965	2.3220	2.3228
C_3H_6	-117.0454	-413.2508	70.6974	-79.2907	2.3226	2.3227
$i\text{-}C_4H_8$	-156.0745	-601.5098	119.4861	-105.7040	2.3227	2.3228
C_3H_4	-115.8401	-388.8190	59.7548	-78.3085	2.3240	2.3248
C_2H_2	-76.8097	-228.1600	24.7300	-51.6582	2.3275	2.3265
C_6H_6	-230.6508	-943.3086	203.3607	-155.3515	2.3265	2.3264
C_2	-75.3621	-205.7492	15.0712	-50.5219	2.3296	2.3302
O_2	-149.5115	-410.0164	27.6771	-93.8435	2.3723	2.3721
CO	-112.6971	-309.7730	22.1348	-72.5298	2.3564	2.3559
CO_2	-187.5628	-559.5041	58.4157	120.0562	2.3599	2.3601
H_2CO	-113.8358	-330.2353	30.9690	-73.1669	2.3573	2.3569
CH_2CO	-151.6824	-472.9917	58.5490	-99.1512	2.3463	2.3463
CH_3OH	-115.0196	-351.4679	40.1893	-73.9800	2.3568	2.3569
$(CH_3)_2O$	-154.0298	-529.1215	83.8112	-100.5990	2.3469	2.3469
$(C_2H_5)_2O$	-232.0913	-923.4393	190.2076	-153.3587	2.3392	2.3397
H_2O	-76.0134	-199.0175	9.1950	-47.4388	2.3759	2.3763
N_2	-108.8826	-301.0868	22.8623	-71.1016	2.3470	2.3453
NH_3	-56.1754	-155.1344	11.7185	-36.7448	2.3459	2.3444
N_2H_2	-109.9529	-323.1350	32.4678	-71.7366	2.3476	2.3483
N_2H_4	-111.1507	-343.0782	41.0728	-72.3681	2.3489	2.3476
HCN	-92.8466	-265.1352	23.9108	-61.1872	2.3410	2.3406
CH_2N_2	-147.7922	-467.9334	61.2808	-98.0200	2.3368	2.3369
NH_2CN	-147.8677	-465.9627	59.7849	-97.3599	2.3416	2.3426
CH_3CN	-131.8884	-423.4763	57.8880	-87.8511	2.3339	2.3330
N_2O	-183.5996	-547.6381	57.8640	-118.5573	2.3543	2.3525
$HCNO$	-167.7063	-513.4983	58.9543	-108.3866	2.3537	2.3588

Results extracted from S. Fliszár, M. Foucrault, M.-T. Béraldin, and J. Bridet, *Can. J. Chem.*, **59**, 1074 (1981).

that $\gamma \simeq 7/3$, within $\pm 2\%$. The reason is apparent from Eq. 5.16: indeed, when applied to atoms, it indicates that $E_{\mathrm{HF}} = (1/\gamma)Z(\partial E_{\mathrm{HF}}/\partial Z)$ and, using also Eqs. 5.8 and 5.11 (with $V_{\mathrm{nn}} = 0$), that $1/\gamma = K^{\mathrm{at}}$. Consequently, we can rewrite Eq. 5.17 as follows for isolated atoms with atomic orbital energies $\varepsilon_i^{\mathrm{orb}}$.

$$\left(3 - \frac{1}{K_k^{\mathrm{at}}}\right) E_k(\text{free atom}) = \sum_i v_i \varepsilon_i^{\mathrm{orb}}. \tag{5.19}$$

So far, the molecular γ is loosely interpreted as a quantity resembling $1/K_k^{\mathrm{at}}$. Assuming temporarily $\gamma = 7/3$, which is the characteristic homogeneity of both Thomas–Fermi and local density functional theory[22], Eq. 5.17 yields the Ruedenberg approximation[23]

$$E = \frac{3}{2} \sum_i v_i \varepsilon_i^{\mathrm{orb}}$$

thus providing an interesting basis for the understanding of approximate energy formulas of the form $E = \text{constant} \times \Sigma_i v_i \varepsilon_i^{\mathrm{orb}}$ which are implied in Hückel and extended Hückel descriptions of molecular energies. Moreover, Eq. 5.18 yields the original Politzer equation[24]

$$E = \frac{3}{7}(V_{\mathrm{ne}} + 2V_{\mathrm{nn}}).$$

This formula represents a valid first approximation. Its usefulness depends, of course, on what is done with it. For our projected application, this approximation turns out to be sufficient. However, although γ is in all cases near the Thomas–Fermi 7/3 limit[13,24], it is clear that its value is not strictly constant[14,21]. The Ruedenberg and Politzer approximations should therefore not be used abusively in problems where a postulated strict constancy of the $E/(V_{\mathrm{ne}} + 2V_{\mathrm{nn}})$ ratio plays a crucial role. Failure to recognize this may cause some disappointments, particularly in tests of the Ruedenberg approximation[25].

So far we have dealt with Hartree–Fock atoms and molecules and have examined selected γ values deduced in this approach. It seems reasonable to anticipate similar results also for real systems although, evidently, slight numerical differences would be expected between exact and Hartree–Fock values of γ—a problem which is of no immediate concern. From here on we assume that γ is defined for real systems by using exact energies instead of Hartree–Fock values in Eqs. 5.16 and 5.18. With these premises, we can now proceed with a closer examination of γ and introduce the defining equation

$$E_k(\text{mol}) = K_k^{\mathrm{mol}} V(k, \text{mol}) \tag{5.20}$$

for the energy $E_k(\text{mol})$ of an atom k being part of a molecule. It is important to stress that at this stage no *a priori* assumption regarding the possible value(s) of the K_k^{mol} factors of the individual atoms is made: these factors are free to assume whatever value is required for satisfying the definition of $E_k(\text{mol})$. With Eq. 5.20 the total molecular energy becomes[13,14]

$$E = \sum_k K_k^{\text{mol}} V(k, \text{mol}) \tag{5.21}$$

because, evidently, $E = \Sigma_k E_k(\text{mol})$. Noting that

$$K_{\text{Av}}^{\text{mol}} = \frac{\sum_k K_k^{\text{mol}} V(k, \text{mol})}{\sum_k V(k, \text{mol})} \tag{5.22}$$

is just a weighted average of the individual K_k^{mol}'s, it follows that[14]

$$E = K_{\text{Av}}^{\text{mol}} \sum_k V(k, \text{mol}) = K_{\text{Av}}^{\text{mol}}(V_{\text{ne}} + 2V_{\text{nn}}) \tag{5.23}$$

Consequently, comparing this result with Eq. 5.18, it appears that

$$\gamma = \frac{1}{K_{\text{Av}}^{\text{mol}}}. \tag{5.24}$$

Equation 5.17 can therefore be written as follows, in its final form for numerical applications[21],

$$\left(3 - \frac{1}{K_{\text{Av}}^{\text{mol}}}\right) E = \sum_i v_i \varepsilon_i^{\text{orb}}. \tag{5.25}$$

The γ parameter is now clearly identified in terms of the K_k^{mol} coefficients of Eq. 5.20 which represent, in a way, the molecular counterparts of the atomic K_k^{at} coefficients appearing in Eq. 5.8. In postulating Eq. 5.21 on grounds of analogies with Eq. 5.8, namely in that total energies are expressed in terms of potentials at the nuclei, Politzer has also assumed a constant K_k^{mol} value for each individual atomic species. The attempt was a successful one at the level of Hartree–Fock energy calculations and consisted essentially of a fit of $\Sigma_k K_k^{\text{mol}} V(k, \text{mol})$ sums with SCF total energies[12-14]. The assumption of a constant K_k^{mol} for each type of atom still merits scrutiny in a more detailed fashion.

Indeed, K_k^{mol} multiplies a total potential energy which consists of two distinct contributions, i.e., (see Eq. 5.9) a nuclear–electronic and a nuclear–nuclear part. It may be regarded as questionable that the same factor should apply to these two distinct parts. The answer is that one can always consider two independent multipliers, K_k^{el} and K_k^{nucl}, so that

$$E_k(\text{mol}) = K_k^{\text{el}} V_{\text{ne}}(k, \text{mol}) + K_k^{\text{nucl}} Z_k \sum_{l \neq k} \frac{Z_l}{r_{lk}} \tag{5.26}$$

In doing so, Eq. 5.20 retains its full validity provided that[21]

$$K_k^{\text{mol}} = \left[K_k^{\text{el}} V_{\text{ne}}(k, \text{mol}) + K_k^{\text{nucl}} Z_k \sum_{l \neq k} \frac{Z_l}{r_{lk}} \right] \left(\frac{1}{V(k, \text{mol})} \right) \tag{5.27}$$

which is simply the weighted average of K_k^{el} (for the nuclear–electronic part) and K_k^{nucl} (for the nuclear–nuclear part). A number of molecules containing carbon, hydrogen, nitrogen, and oxygen were studied in order to determine the appropriate independent coefficients K_k^{el} and K_k^{nucl} for each of these atoms. This was done by separating the nuclear–electronic from the nuclear–nuclear contributions for each chemical species, using Eq. 5.26, and adding them separately to give sub-sums which collect only the contributions of atoms of the same chemical species. The individual coefficients derived in this manner from Hartree–Fock total and potential energies indicate that, within the precision of this type of analysis, $K_k^{el} = K_k^{nucl}$. These results[21] settle an important question, namely that of the constancy of the K_k^{mol} factor introduced in the defining equation 5.20. Indeed, had it been $K_k^{el} \neq K_k^{nucl}$, then K_k^{mol} would have changed to some extent from molecule to molecule, depending on the relative weights of $V_{ne}(k, mol)$ and $Z_k \Sigma_{l \neq k} Z_l / r_{lk}$ (see Eq. 5.27).

At this stage we have collected all the appropriate information concerning relationships between energy components in atoms and in molecules at equilibrium and reduced the problem of expressing molecular energies to a purely electrostatic one, involving only nuclear–electronic and nuclear–nuclear interactions. Although, of course, much could be added on related topics, namely on exact and local density functional theory, homogeneity of total $vs.$ electronic energy, and the like, which are the object of current investigations[26], we shall now proceed with the projected study of atomization energies in terms of electron distributions.

5. BINDING OF AN ATOM IN A MOLECULE

We can now examine the energy difference

$$\Delta E_k = E_k(\text{free atom}) - E_k(\text{mol}) \tag{5.28}$$

between a free and a bonded atom. ΔE_k is clearly a measure for the process of a free atom becoming part of a molecule and contains a portion of the molecular binding energy. The sum

$$\Delta E_a^* = \sum_k \Delta E_k$$

represents, accordingly, the total energy of atomization at 0 K of a molecule in its hypothetical vibrationless state since, by definition, $\Delta E_a^* = \Sigma_k E_k(\text{free atom}) - E_{mol}$ and $E_{mol} = \Sigma_k E_k(\text{mol})$. Introducing at this stage the concept of "chemical bond", we consider that the ΔE_a^* energy includes a contribution, $\Delta E_{nb}^* = 0 - E_{nb}^*$, required to annihilate all the interactions between non-bonded atoms (in the usual "chemical" acceptation of this term) and write[27]

$$\Delta E_a^* = \Delta E_a^{*\,\text{bonds}} - E_{nb}^*. \tag{5.29}$$

Regarding the nonbonded interactions, Del Re has shown that a valid approximation in σ systems is Coulombic in nature[28], i.e.,

$$E_{nb}^* = \frac{1}{2} \sum_{k,l}^{nb} \frac{q_k q_l}{r_{kl}} \tag{5.30}$$

where q_k and q_l are the net atomic charges of nonbonded atom pairs at a distance r_{kl}. Numerical evaluations indicate that $|E_{nb}^*| \ll \Delta E_a^*$, typically $\sim 0.01 - 0.05\%$ of the total energy of atomization for molecules like ethane, cyclohexane, adamantane, di-n-propylether, etc. Coulomb interactions between nonbonded atoms cannot, indeed, be made responsible for energy differences between isomers. For example, the difference in nonbonded energy between normal pentane and neopentane is only ~ 0.1 kcal/mol favoring, as expected, the normal isomer, whereas neopentane is actually the more stable form by ~ 4 kcal/mol in total atomization energy. Similarly, di-n-propylether and diisopropylether differ from one another by ~ 0.2 kcal/mol in nonbonded energy (in favor of the normal isomer), although the branched form is the more stable in ΔE_a^* energy, by ~ 6.8 kcal/mol. The main quantity responsible for the structure-related energy effects is thus $\Delta E_a^{* \, \text{bonds}}$, whose behavior governs the major part, by far, of the properties determining ΔE_a^*.

Defining now the "bond energy" ε_{ij} as the portion of $\Delta E_a^{* \, \text{bonds}}$ associated with the bond formed by atoms i and j, we treat the bonded part of ΔE_a^* as a sum of energy terms ε_{ij} referring to the individual bonds ij, i.e.,

$$\Delta E_a^* = \sum_{i>j} \varepsilon_{ij} - E_{nb}^*. \tag{5.31}$$

Using the exact quantum mechanical definition

$$\Delta E_a^* = \sum_k \langle \psi_{at}^k | H_{at}^k | \psi_{at}^k \rangle - \langle \psi_{mol} | H_{mol} | \psi_{mol} \rangle$$

where H_{at}^k and H_{mol} are the appropriate Hamiltonians and ψ_{at}^k and ψ_{mol} the corresponding ground-state wave functions, we calculate the derivative $\partial \Delta E_a^* / \partial Z_k$ with respect to the nuclear charge Z_k of the kth atom in the molecule leaving the internuclear distances and the number of electrons unchanged. It follows from the Hellmann–Feynman theorem that

$$\partial \Delta E_a^* / \partial Z_k = \left\langle \psi_{at}^k \left| \frac{\partial H_{at}^k}{\partial Z_k} \right| \psi_{at}^k \right\rangle - \left\langle \psi_{mol} \left| \frac{\partial H_{mol}}{\partial Z_k} \right| \psi_{mol} \right\rangle$$

$$= \left\langle \psi_{at}^k \left| -\sum_i r_{ik}^{-1} \right| \psi_{at}^k \right\rangle - \left\langle \psi_{mol} \left| -\sum_i r_{ik}^{-1} + \sum_{l \neq k} \frac{Z_l}{r_{lk}} \right| \psi_{mol} \right\rangle$$

The first sum over i runs over all electrons in atom k with wave function ψ_{at}^k and the corresponding integral is V_{ne}(free atom k)$/Z_k$. The second sum over i runs over all electrons in the molecule and the corresponding integral is $V_{ne}(k, \text{mol})/Z_k$. Finally, l runs over all nuclei but k, and r_{lk} is the distance between nuclei k and l. It follows that

$$V(k, \text{mol}) = V_{\text{ne}}(\text{free atom } k) - Z_k \frac{\partial \Delta E_{\text{a}}^*}{\partial Z_k} \qquad (5.32)$$

with $V(k, \text{mol})$ as indicated in Eq. 5.9. The only terms contributing to the derivative $\partial \Delta E_{\text{a}}^*/\partial Z_k$ are those involving atom k, namely, its bonded interactions with atom(s) j and nonbonded interactions with all other atoms, giving $\partial \Delta E_{\text{a}}^*/\partial Z_k = \Sigma_j \partial \varepsilon_{kj}/\partial Z_k - \partial E_{\text{nb}}^*/\partial Z_k$. Neglecting temporarily the nonbonded contributions, we deduce from Eq. 5.32 that [14,27]

$$V(k, \text{mol}) = V_{\text{ne}}(\text{free atom } k) - Z_k \sum_j \frac{\partial \varepsilon_{kj}}{\partial Z_k} \qquad (5.33)$$

This approximation for the "true" $V(k, \text{mol})$ potential energy is certainly valid when $|E_{\text{nb}}^*| \ll \Delta E_{\text{a}}^{* \text{bonds}}$. More appropriately, however, we regard this expression as a description of that portion of the total $V(k, \text{mol})$ which refers precisely to the bonded part of ΔE_{a}^*. We can thus safely proceed by using Eq. 5.33, just bearing in mind that the quantities derived therefrom refer to molecules stripped of their nonbonded interactions. In this sense, the validity of Eq. 5.33 is determined only by the validity of apportioning ΔE_{a}^* into bonded and nonbonded terms, i.e., ultimately, by the very existence of "chemical bonds".* Equation 5.33 is instrumental in numerical evaluations of the $\partial \varepsilon_{kj}/\partial Z_k$ derivatives, a task which is not free from difficulties. Fortunately, with the selection of a convenient set of "model" bonds (e.g., the CC and CH bonds of ethane), we only need to calculate a limited number of "reference" derivatives, $\partial \varepsilon_{ij}^\circ/\partial Z_i$, which correspond to the ε_{ij}° energies of the reference bonds. Selected results are given in Table 5.4.

By means of Eq. 5.33 it is now possible to derive an instructive expression for the energy difference ΔE_k (Eq. 5.28) between an isolated and a bonded atom, using also the appropriate equations 5.8 and 5.20. The result [14,27]

$$\Delta E_k = K_k^{\text{mol}} Z_k \sum_j \frac{\partial \varepsilon_{kj}}{\partial Z_k} + (K_k^{\text{at}} - K_k^{\text{mol}}) V_{\text{ne}}(\text{free atom } k) \qquad (5.34)$$

stresses the role of local binding properties in determining ΔE_k. This equation translates the concept of a molecule viewed as a collection of chemical bonds into a description in terms of "atoms in a molecule". Namely, it appears that besides the small nonbonded contribution which evidently depends on the whole of the molecule, ΔE_k is primarily related both to the type and to the number of bonds formed by atom k. For numerical evaluations of ΔE_k, we calculate the required atomic K_k^{at} values by means of Eq. 5.8 or 5.19 from the appropriate Hartree–Fock results, as well as the corresponding K_k^{mol} parameters, either from Eq. 5.23 or 5.25 using Hartree–Fock results derived for the homonuclear diatomics (in which case $K_{\text{Av}}^{\text{mol}} = K_k^{\text{mol}}$) or, else,

---

*Note that for any bonded atom pair we speak in all cases of *one* bond only, whether this bond is a single one (in conventional language), like the CC bond in ethane, or a multiple bond (like the CC bonds in ethylene, acetylene, benzene, for example).

TABLE 5.4. $\partial\varepsilon_{ij}^{\circ}/\partial Z_i$ Derivatives for Selected Reference Bonds

Molecule	Bond	$\partial\varepsilon_{ij}^{\circ}/\partial Z_i$ (a.u.)
Ethane	CH	$\partial\varepsilon_{CH}^{\circ}/\partial Z_C =$ 0.027
	CH	$\partial\varepsilon_{CH}^{\circ}/\partial Z_H =$ 0.153
	CC	$\partial\varepsilon_{CC}^{\circ}/\partial Z_C =$ 0.012
Ethylene	CC	$\partial\varepsilon_{CC}^{\circ}/\partial Z_C =$ 0.0305
Diethylether	CO	$\partial\varepsilon_{CO}^{\circ}/\partial Z_C =$ -0.049
	CO	$\partial\varepsilon_{CO}^{\circ}/\partial Z_O =$ 0.036
Acetone	CO	$\partial\varepsilon_{CO}^{\circ}/\partial Z_C =$ -0.077
	CO	$\partial\varepsilon_{CO}^{\circ}/\partial Z_O =$ 0.071

Calculated from Eq. 5.33, on the basis of HF results. Details are given in Ref. 27, as well as in: S. Fliszár and M.-T. Béraldin, *Can. J. Chem.*, **60**, 792 (1982); M.-T. Béraldin and S. Fliszár, *Can. J. Chem.*, **61**, 197 (1983). Additional examples illustrating the use of Eqs. 5.33 and 5.35 are offered in Chapter 8.6.

TABLE 5.5. Selected K_k^{at} and K_k^{mol} Values

Atom	K_k^{at}	K_k^{mol}
H	0.5000	0.5000
B	0.4311	0.4321
C	0.4276	0.4287
N	0.4239	0.4260
O	0.4201	0.4213
F	0.4165	0.4169

The K_k^{mol} data follow from total and potential energies (Ref. 21, Eq. 5.21) and/or total and orbital energies of diatomics (Eq. 5.25) given by P.E. Cade and W. Huo, *At. Data Nucl. Data Tables*, **12**, 415 (1973); P.E. Cade and A.C. Wahl, *ibid.*, **13**, 339 (1974).

from multiple regression analyses[21] involving SCF inputs for Eq. 5.21. Selected results are presented in Table 5.5. For the carbon and hydrogen atoms of ethane, Eq. 5.34 gives $\Delta E_C = 0.3366$ and $\Delta E_H = 0.0765$ a.u., while for the diethylether and acetone oxygen atoms ΔE_O is 0.457 and 0.454 a.u., respectively. The positive values reflect the fact that in these cases the bonded atoms are more stable than the isolated ones. (This type of result is, however, not a general one, as indicated by the acetone carbonyl-C atom, with $\Delta E_C \simeq -0.04$ a.u.). These results featuring ΔE_k illustrate the "atoms in the molecule" aspect of the present theory.

The chemical bonds themselves are also well described by Eq. 5.34. Their energies are deduced by the following decomposition of ΔE_k among the bonds formed by atom k. First, the "extraction" from the host molecule of an atom i forming v_i bonds requires an energy $(K_i^{at} - K_i^{mol})V_{ne}$(free atom i)/v_i for each bond, meaning that in the suppression of an ij bond this type of contribution must be counted once for both the i and j atoms. In addition, this atomization requires an energy $K_i^{mol}Z_i\partial\varepsilon_{ij}/\partial Z_i$ for each bond formed by i, meaning that the cleavage of an ij bond involves this type of contribution for each bonded partner. Consequently, the portion of the total atomization energy associated with the ij bond is[27]

$$\varepsilon_{ij} = K_i^{mol}Z_i\left(\frac{\partial\varepsilon_{ij}}{\partial Z_i}\right) + K_j^{mol}Z_j\left(\frac{\partial\varepsilon_{ij}}{\partial Z_j}\right) + (K_i^{at} - K_i^{mol})V_{ne}(\text{free atom } i)\left(\frac{1}{v_i}\right)$$

$$+ (K_j^{at} - K_j^{mol})V_{ne}(\text{free atom } j)\left(\frac{1}{v_j}\right). \tag{5.35}$$

TABLE 5.6. Energy Terms of Selected Reference Bonds, kcal/mol

		ε_{ij}°	
Reference Molecule	Bond	Theor. (Eq. 5.35)	"Best"
Ethane	CC	69.3	69.633
	CH	106.9	106.806
Ethylene	CC	139.2	139.27
Diethylether	CO	80	79.78
Acetone	CO	181	179.40

The appropriate V_{ne}(free atom i) values were derived from Eq. 5.8 using the atomic energies $E_H = -1/2$, $E_C = -37.8558$, and $E_O = -75.1102$ a.u., which were taken as the sums of the ionization potentials (with negative signs) given in: C.E. Moore, Nat. Stand. Ref. Data Ser. (US Nat. Bur. Stand.), NSRDS–NBS **34** (1970). The conversion factor 1 a.u. = 27.2106 eV was used.

This energy formula, which is the explicit "chemical bond" counterpart of Eq. 5.34, illustrates clearly the equivalence of the models describing molecules in terms of atomic-like contributions or, alternatively, in terms of chemical bonds. Using the appropriate derivatives given in Table 5.4 and the parameters of Table 5.5, we deduce the corresponding theoretical "reference" ε_{ij}° energies by means of Eq. 5.35. Of course, any error in their evaluation leads to systematic deviations which are magnified by the number of their occurrence in a molecule. An empirical refinement, using experimental atomization energies, yields the "best" ε_{ij}°'s which are compared in Table 5.6 with their theoretical counterparts. (We postpone temporarily the question of how the ethane CC and CH values relate to the customary empirical ones, ~ 82 and ~ 105 kcal/mol, respectively.)

With Eq. 5.35 we have accomplished part of the intended energy analysis of a molecule in terms of "local" contributions associated with individual chemical bonds. Although, of course, numerical values derived in this fashion are interesting, the major merit of Eq. 5.35 (and of its "atomic" counterpart, Eq. 5.34) is to draw attention on simple important facts of fundamental importance, which shall now be put in a clear perspective.

6. NON-TRANSFERABILITY OF BOND ENERGY TERMS

Equations 5.34 and 5.35 are two alternate descriptions of the same reality. Their equivalence is best illustrated by taking ethane as an example. Using the results given in Tables 5.5 and 5.6, we find for this molecule that the calculated result $2\Delta E_C + 6\Delta E_H = \varepsilon_{CC} + 6\varepsilon_{CH} = 710.6$ kcal/mol is reasonably close to the experimental one, 710.54 kcal/mol. Assuming now for a moment that the derivatives $\partial\varepsilon_{kj}/\partial Z_k$ of Table 5.5 can be treated as constants, it results from Eq. 5.34 that $\Delta E_C = 0.2980$ for a CH_2 carbon attached to 2 other C atoms, 0.2594 for a CH carbon attached to 3 other C atoms, and 0.1411 a.u. for the CH_2 carbon atom of diethylether. With these values, we observe that the result deduced for adamantane, $\Delta E_a^{*\,\text{bonds}} = 6\Delta E_C(\text{sec}) + 4\Delta E_C(\text{tert}) + 16\Delta E_H = 12\varepsilon_{CC} + 16\varepsilon_{CH} \simeq 2541$ kcal/mol, is in error by ~ 147 kcal/mol with respect to the experimental value, 2688.05 kcal/mol. Similarly, the result calculated for diethylether is in error by ~ 23 kcal/mol with respect to its experimental counterpart, 1389.87 kcal/mol.

These examples raise the obvious question about the origin of the discrepancies between observed and calculated atomization energies which are known to plague simple bond additivity schemes. A popular way of dealing with this problem is to invoke "steric effects", corresponding to given topological situations, which are adequately parametrized in order to restore some sort of agreement between "expected" and observed energies. The physical origin of this type of "interactions" is, however, largely left to the imagination, a fact which should not be hidden by the usually good numerical quality and practical usefulness of empirical correlations constructed in terms of "steric effects". It is clear, on the other hand, that Coulomb-type interactions (Eq. 5.30) involving nonbonded atoms cannot be invoked for filling the gap between simple bond-additivity and experimental results, not only because of the small weight of this type of contribution, but also because their variations are too often in the wrong direction. With this in mind, we examine now the bonded contributions.

First of all, we note that any sum $\Sigma\Delta E_k = \Sigma\varepsilon_{ij}$ constructed, as we did, from a fixed set of ΔE_k or ε_{ij} values is a clear representation of exact additivity. The clue to the correct meaning of this sum lies in the precise definition of the derivatives $\partial\varepsilon_{ij}/\partial Z_k$ which enter the calculation of the ΔE_k and ε_{ij} terms. These derivatives are, indeed, bound to the same conditions which apply in the present use of the Hellmann–Feynman theorem; namely, they are carried out leaving the internuclear distances and the number of electrons unchanged. The appropriate derivatives should, therefore, be calculated in each case of interest for the specified kj bond to which they refer, i.e., for a specified situation described by the kj internuclear distance and the electron distributions about the atoms involved. Instead, with the selection of a fixed set of $\partial\varepsilon_{kj}/\partial Z_k$ values, we end up using "model" bonds or "model" atoms (e.g., those of ethane in the examples given above), disregarding possible changes in internuclear distances and electron populations. Now, the simple sum of constant bond (or atomic) terms, implying a fixed set of $\partial\varepsilon_{kj}/\partial Z_k$ derivatives and, hence, invariant "local" electron populations, cannot (as a rule) describe an electroneutral molecule. Indeed, if *constant* electron populations $N_A(\neq Z_A)$, $N_B(\neq Z_B)$, ... are associated with all individual atoms A, B, ... of an electroneutral molecule, any other nonisomeric molecule constructed from the same atoms with the same charges would not satisfy the charge normalization condition[27]. For example, if in ethane the C and H net charges are 0.0351 and -0.0117 e, respectively, the $C_{10}H_{16}$ hydrocarbon adamantane constructed from these atoms would be electron deficient by 0.1638 electron.

Hence, unless one denies entirely the existence of charge transfers within molecules, this argument suffices to put any additivity scheme involving fixed $\partial \varepsilon_{kj}/\partial Z_k$ derivatives on the disabled list. The discrepancies between observed and exactly "additive" energies cannot be explained on "steric" grounds alone, at least not if these are viewed as some sort of distant interactions between nonbonded atoms. Rather, in the search for a satisfactory expression for the bonded contributions, we must explicitly include a charge dependence in the ε_{ij} (or ΔE_k) terms and restore electroneutrality. In doing so we consider, as succinctly stated by Platt[29], that "a theory of chemistry and the chemical bond is primarily a theory of electron density".

7. CHARGE DEPENDENCE OF CHEMICAL BINDING

The bond energy terms ε_{ij} deduced from a well-defined set of $\partial \varepsilon_{ij}/\partial Z_i$ values are, from here on, designated by ε_{ij}°. The difference

$$-E(\text{charge}) = \Delta E_a^{*\,\text{bonds}} - \sum \varepsilon_{ij}^\circ \qquad (5.36)$$

measures the effect of using in each case the appropriate charge distributions, instead of frozen ones. (The sign is determined by that of the $\Delta E_a^{*\,\text{bonds}}$ and $\sum \varepsilon_{ij}^\circ$ energies which are for "atoms minus molecule"). The calculation of $E(\text{charge})$ is, in fact, one that relates the changes from ε_{ij}° to ε_{ij} to changes in the electronic structures of atoms i and j. For small perturbations, the latter are considered to occur in the valence shells, leaving the core regions unaltered. The following calculations are carried out in the spirit of the Politzer–Parr electron partitioning into core and valence regions (see Sect. 3), which is also meaningful in molecules. The nuclear charges Z_i^{eff} and Z_j^{eff} are effective charges, e.g., $Z^{\text{eff}}(\text{carbon}) \simeq 4$ a.u. and $Z^{\text{eff}}(\text{oxygen}) \simeq 6$ a.u. Similarly, electron densities (ρ), populations (N), and energies refer to the valence shells.

$E(\text{charge})$ is most conveniently derived from the change $\Delta V_{\text{ne}}(\text{charge})$ in nuclear-electronic potential energy accompanying the appropriate charge normalization. This $\Delta V_{\text{ne}}(\text{charge})$ correction, of course, concerns only the interactions between bonded atoms. When added to the sum $\sum_k V(k, \text{mol})$ obtained by using Eq. 5.33, the result differs from the exact one only by the omitted nonbonded contributions. Since the exact $\sum_k V(k, \text{mol})$ sum and the one derived from Eq. 5.33 both are to be multiplied by $K_{\text{Av}}^{\text{mol}}$ to give the corresponding molecular energies (i.e., respectively, the total energy and that of non-charge-normalized molecules stripped of their nonbonded contributions), it appears safe to write

$$E(\text{charge}) = K_{\text{Av}}^{\text{mol}} \Delta V_{\text{ne}}(\text{charge}). \qquad (5.37)$$

The problem of calculating $E(\text{charge})$ thus reduces to a calculation of nuclear-electronic potential energies. The contribution to $\Delta V_{\text{ne}}(\text{charge})$ involving Z_i^{eff} consists, first, of its interactions with the electrons of atom i

$$-Z_i^{\text{eff}} \int_{\tau_i} \frac{1}{|\mathbf{r} - \mathbf{r}_i|} \rho_i(\mathbf{r}) d\mathbf{r}$$

where the integration is carried out over the volume τ_i containing the N_i electrons allocated to atom i, and, second, of a part

$$-Z_i^{\text{eff}} \int_{\tau_j} \frac{1}{|\mathbf{r} - \mathbf{r}_i|} \rho_j(\mathbf{r}) d\mathbf{r}$$

referring to the interaction with the N_j electrons of each atom j bonded to i. The above integrals are conveniently written as

$$\int_{\tau_i} \frac{1}{|\mathbf{r} - \mathbf{r}_i|} \rho_i(\mathbf{r}) d\mathbf{r} = N_i \langle r_i^{-1} \rangle$$

and

$$\int_{\tau_j} \frac{1}{|\mathbf{r} - \mathbf{r}_i|} \rho_j(\mathbf{r}) d\mathbf{r} = N_j \langle r_{ij}^{-1} \rangle$$

where $\langle r_i^{-1} \rangle$ and $\langle r_{ij}^{-1} \rangle$ are the average inverse distances from Z_i^{eff} to N_i and N_j, respectively. Similar expressions are written for the reference molecule with atomic electron populations N_i°, N_j° and average inverse distances $\langle r_i^{-1} \rangle^\circ$, $\langle r_{ij}^{-1} \rangle^\circ$. In this manner we avoid the explicit calculation of the electron densities ρ and, moreover, postpone the precise definition of the appropriate N_i's, i.e., the problem of electron partitioning. The molecular ΔV_{ne}(charge) correction is then simply given by Eq. 5.38.

$$\Delta V_{\text{ne}}(\text{charge}) = -\sum_i Z_i^{\text{eff}} \left[N_i \langle r_i^{-1} \rangle - N_i^\circ \langle r_i^{-1} \rangle^\circ \right. $$
$$\left. + \sum_j (N_j \langle r_{ij}^{-1} \rangle - N_j^\circ \langle r_{ij}^{-1} \rangle^\circ) \right]. \tag{5.38}$$

What this equation says is simply that the addition of a small amount of electronic charge to an atom i changes its "own" atomic nuclear–electronic interaction energy, i.e., that involving nucleus i, and, moreover, also changes the nuclear–electronic interactions with the nuclei of all atoms j bonded to i. The effects involving all the other atoms of the molecule are, of course, included in the evaluation of nonbonded interaction energies, which is a separate calculation.

At this stage we introduce the approximation

$$\langle r_{ij}^{-1} \rangle \simeq \langle r_{ij}^{-1} \rangle^\circ \tag{5.39}$$

which expresses the simplifying hypothesis that small perturbations in the valence-shell electron populations do not affect their shape, i.e., their center of charge. For nearly spherical atomic charge clouds, we approximate $\langle r_{ij}^{-1} \rangle^\circ$ by the inverse of the internuclear distance $(r_{ij}^{-1})_{\text{nn}}$. The $(r_{ij}^{-1})_{\text{nn}}$'s are temporarily kept constant in comparisons between bonds of similar nature (e.g., the CC or the CH bonds in saturated hydrocarbons) because we are presently concerned only with the charge effects on the ε_{ij}'s. Defining now

$$N_i = N_i^\circ + \Delta N_i \tag{5.40}$$

it follows from Eqs. 5.37–5.40 that

$$E(\text{charge}) = -K_{\text{Av}}^{\text{mol}} \sum_i Z_i^{\text{eff}} \left(N_i \langle r_i^{-1} \rangle - N_i^\circ \langle r_i^{-1} \rangle^\circ + \sum_j \langle r_{ij}^{-1} \rangle^\circ \Delta N_j \right). \tag{5.41}$$

As for the difference $N_i \langle r_i^{-1} \rangle - N_i^\circ \langle r_i^{-1} \rangle^\circ$ appearing in Eq. 5.41, first we expand $N_i \langle r_i^{-1} \rangle$ in a Taylor series

$$N_i \langle r_i^{-1} \rangle = N_i^\circ \langle r_i^{-1} \rangle^\circ + \left(\frac{\partial N_i \langle r_i^{-1} \rangle}{\partial N_i} \right)^\circ \Delta N_i + \frac{1}{2!} \left(\frac{\partial^2 N_i \langle r_i^{-1} \rangle}{\partial N_i^2} \right)^\circ (\Delta N_i)^2 + \cdots \tag{5.42}$$

and, second, define the energy

$$E_i^{\text{vs}} = -K_i^{\text{mol}} Z_i^{\text{eff}} N_i \langle r_i^{-1} \rangle \tag{5.43}$$

of atom i in its valence state, in the current acceptation of this term, chosen so as to have, as nearly as possible, the same interaction between the electrons of the atom, as occurs when the atom is part of a molecule. The valence state is considered as being formed from a molecule by removing all the other atoms without allowing any electronic rearrangement in the given atom and differs, hence, from the energy given by Eq. 5.20 by the non-inclusion of the electronic and nuclear interactions due to the other atoms of the molecule. Taking now the successive derivatives of E_i^{vs} (Eq. 5.43) evaluated for $N_i = N_i^\circ$, i.e.,

$$\left(\frac{\partial E_i^{\text{vs}}}{\partial N_i} \right)^\circ = -K_i^{\text{mol}} Z_i^{\text{eff}} \left(\frac{\partial N_i \langle r_i^{-1} \rangle}{\partial N_i} \right)^\circ$$

$$\left(\frac{\partial^2 E_i^{\text{vs}}}{\partial N_i^2} \right)^\circ = -K_i^{\text{mol}} Z_i^{\text{eff}} \left(\frac{\partial^2 N_i \langle r_i^{-1} \rangle}{\partial N_i^2} \right)^\circ, \text{ etc.,}$$

we obtain from Eq. 5.42 that

$$N_i \langle r_i^{-1} \rangle - N_i^\circ \langle r_i^{-1} \rangle^\circ = \frac{-1}{K_i^{\text{mol}} Z_i^{\text{eff}}} \left[\left(\frac{\partial E_i^{\text{vs}}}{\partial N_i} \right)^\circ \Delta N_i + \frac{1}{2!} \left(\frac{\partial^2 E_i^{\text{vs}}}{\partial N_i^2} \right)^\circ (\Delta N_i)^2 + \cdots \right]$$

and, from Eq. 5.41, that

$$E(\text{charge}) = K_{\text{Av}}^{\text{mol}} \sum_i \left[\frac{1}{K_i^{\text{mol}}} \left(\frac{\partial E_i^{\text{vs}}}{\partial N_i} \right)^\circ \Delta N_i + \frac{1}{2! K_i^{\text{mol}}} \left(\frac{\partial^2 E_i^{\text{vs}}}{\partial N_i^2} \right)^\circ (\Delta N_i)^2 \right.$$

$$\left. + \cdots - Z_i^{\text{eff}} \sum_j \langle r_{ij}^{-1} \rangle^\circ \Delta N_j \right]. \tag{5.44}$$

Finally, when the result is expressed in terms of net (i.e., nuclear minus electronic) charges

$$\Delta q = -\Delta N$$

it follows from Eq. 5.36 that

$$\Delta E_a^{*\,\text{bonds}} = \sum \varepsilon_{ij}^\circ + K_{\text{Av}}^{\text{mol}} \sum_i \left[\frac{1}{K_i^{\text{mol}}} \left(\frac{\partial E_i^{\text{vs}}}{\partial N_i} \right)^\circ \Delta q_i + \cdots - Z_i^{\text{eff}} \sum_j \langle r_{ij}^{-1} \rangle^\circ \Delta q_j \right].$$

(5.45)

This equation contains the full information about the effects of atomic charges in determining $\Delta E_a^{*\,\text{bond}}$ energies, showing that isomers or conformers differing in their electron distributions differ in the overall stability of their chemical bonds. The charge-dependent part provides, thus, an interpretation of "steric effects" in terms of electron distributions, a topic which shall be discussed further below. Note that molecular electroneutrality is now restored with the use of the appropriate Δq's.

A more enlightening result can be derived from Eq. 5.45 by observing that the sum over all atoms i can be rearranged into an equivalent form involving individual bond contributions, giving for each ij bond[27]

$$\varepsilon_{ij} = \varepsilon_{ij}^\circ + a_{ij}\Delta q_i + a_{ji}\Delta q_j$$

(5.46)

where

$$a_{ij} = \frac{1}{v_i} \frac{K_{\text{Av}}^{\text{mol}}}{K_i^{\text{mol}}} \left[\left(\frac{\partial E_i^{\text{vs}}}{\partial N_i} \right)^\circ - \frac{1}{2} \left(\frac{\partial^2 E_i^{\text{vs}}}{\partial N_i^2} \right)^\circ \Delta q_i \right] - K_{\text{Av}}^{\text{mol}} Z_j^{\text{eff}} \langle r_{ij}^{-1} \rangle^\circ$$

(5.47)

with v_i = number of atoms attached to atom i. The sum over all the bonds is then

$$\Delta E_a^{*\,\text{bonds}} = \sum \varepsilon_{ij}^\circ + \sum_i \sum_j a_{ij}\Delta q_i.$$

(5.48)

The proof follows from the sum $\Delta E_a^{*\,\text{bonds}} = \Sigma\varepsilon_{ij}$ which yields Eq. 5.45. Equations 5.46–5.48 are the most convenient working formulas in numerical evaluations of $\Delta E_a^{*\,\text{bonds}}$ energies and for the discussion of individual bond contributions in relation to charge effects. The approximations involved in these equations are mild ones. Third-order derivatives $\partial^3 E_i^{\text{vs}}/\partial N_i^3$ need not be retained as the second-order derivatives are virtually constant, at least in the range of Δq's which are considered. Actually, the second-order term $(1/2)\,(\partial^2 E_i^{\text{vs}}/\partial N_i^2)\Delta q_i$ of Eq. 5.47, not indicated explicitly in Eq. 5.45, can also be neglected in many cases (e.g., for carbon atoms in hydrocarbons), although it must be retained in others (e.g., for carbonyl oxygen atoms whose Δq_O's are relatively important). Similarly, no significant bias is expected from the approximation $\langle r_{ij}^{-1} \rangle \simeq \langle r_{ij}^{-1} \rangle^\circ$ (Eq. 5.39). This can be shown as follows[27].

A complete description of $\Delta E_a^{*\,\text{bonds}}$ should also allow for changes in internuclear distances. Including now the corresponding $V_{nn} - V_{nn}^\circ$ term into the sum $\Sigma_i \Sigma_j Z_j^{\text{eff}}(N_j\langle r_{ij}^{-1}\rangle - N_j^\circ \langle r_{ij}^{-1}\rangle^\circ)$ appearing in Eq. 5.38, we evaluate the new sum

$$\sum_i \sum_j \{Z_i^{\text{eff}} Z_j^{\text{eff}}[(r_{ij}^{-1})_{\text{nn}} - (r_{ij}^{-1})_{\text{nn}}^\circ] - Z_i^{\text{eff}}(N_j\langle r_{ij}^{-1}\rangle - N_j^\circ \langle r_{ij}^{-1}\rangle^\circ)\}$$

by using the definition $Z_j^{\text{eff}} - N_j^\circ = q_j^\circ$ of net atomic charges. Comparison with Eq. 5.45 indicates then that the missing term in this expression is

$$K_{\text{Av}}^{\text{mol}} \sum_i \sum_j \{Z_i^{\text{eff}} Z_j^{\text{eff}}[(r_{ij}^{-1})_{\text{nn}} - (r_{ij}^{-1})_{\text{nn}}^\circ - (\langle r_{ij}^{-1}\rangle - \langle r_{ij}^{-1}\rangle^\circ)]$$
$$+ Z_i^{\text{eff}} q_j^\circ(\langle r_{ij}^{-1}\rangle - \langle r_{ij}^{-1}\rangle^\circ)\}$$

which represents the contribution to $\Delta E_a^{*\text{bonds}}$ due to variations in internuclear distances and to changes of electronic centers of charge. The first part, in square brackets, is obviously 0 for spherically symmetric electron clouds and, more generally, if the centers of electronic charge move along with the nuclei during small changes in internuclear distances. The last part is small because atomic net charges are small in the first place (e.g., $\leqslant 0.0351$ e for C) and because small changes in electron populations are unlikely to modify their center of charge to any significant extent. Finally, a practical way of minimizing errors of this sort rests with the selection of the appropriate $(r_{ij}^{-1})_{\text{nn}}$ when calculating a_{ij}. Indeed, in studies of homologous series, it is usually easy to select the appropriate distances so as to satisfy the cases where the Δq's are largest. Clearly, for small Δq's, slight errors in a_{ij} result in insignificant uncertainties which are usually within experimental error. Under these circumstances, it can be concluded that Eqs. 5.46–5.48 are, indeed, satisfactory approximations, at least for a wide variety of molecules.

From here on, the study of charge effects in determining molecular stabilities consists of numerical applications of Eqs. 5.46–5.48. While, of course, comprehensive calculations for use in comparisons with experimental energies require the detailed knowledge of the appropriate charges, much can be learned without this knowledge, from the study of the a_{ij} parameters.

8. EVALUATION OF THE a_{ij} PARAMETERS

We now direct our attention to the calculation of the a_{ij}'s (Eq. 5.47). The first and second derivatives, $\partial E_i^{\text{vs}}/\partial N_i$ and $\partial^2 E_i^{\text{vs}}/\partial N_i^2$, are conveniently obtained from SCF–Xα theory[30], which offers the advantage of permitting this type of calculations for any desired integer and fractional electron population. It is important, indeed, to realize, and account for, the fact that the $\partial E_i^{\text{vs}}/\partial N_i$ values depend on N_i. The difficulty is that such calculations cannot be performed directly on atoms which are actually part of a molecule, so that we resort to appropriate free atom calculations in order to mimic the behavior of atoms which are in a molecule, but do not experience external interactions due to the presence of the other atoms forming the molecule. For hydrogen we have used the α value (0.686) appropriate for partially negative

H (like that of ethane) and which reproduces correctly its electron affinity. For $N_H = 1.0117$ e (corresponding to $q_C^\circ = 0.0351$ e in ethane), it is found that $\partial E_H/\partial N_H = -0.195$ a.u. For the carbon atoms in saturated hydrocarbons, we have considered, first, that fully optimized *ab initio* studies of hydrocarbons indicate that any gain in electronic charge, with respect to the ethane carbon, occurs at the $2s$ level[31]. Second, GTO($9s5p|6s$) → $[5s3p|3s]$ calculations of methane and ethane, using Dunning's exponents[32] and optimum contraction vectors[31], indicate $2s$ populations of 1.42–1.46 e. Finally, SCF–Xα calculations indicate $\partial E_i/\partial N_i$ values of -20.49, -19.87, and -19.26 eV for $2s$ populations of 1.40, 1.45, and 1.50 e, respectively, by using the $\alpha = 0.75928$ value given by Schwarz[33]. These results suggest that the appropriate $\partial E_C/\partial N_C$ derivative can be reasonably estimated at -0.735 Hartree a.u. (-20 eV). The selection of the appropriate derivative for oxygen presents no difficulties because *ab initio* population analyses clearly show the involvement of $2p$ electrons in this case. At carbonyl-carbon atoms, mainly $2p$ populations are affected by substitution. The results for C and O $2p$ electrons were deduced using the "frozen core" α values given by Schwarz[33], i.e., $\alpha_C = 0.75928$ and $\alpha_O = 0.74447$ (Table 5.7).

Turning now to the other terms appearing in Eq. 5.47, we approximate K_{Av}^{mol} by 3/7 and use the K_i^{mol} results indicated in Table 5.5. The $\langle r_{ij}^{-1} \rangle$ terms are approximated by the inverse internuclear distances, $(r_{ij}^{-1})_{nn}$. The parameters and the final results[27,34] deduced from Eq. 5.47 are given in Table 5.7.

A complication arises when atomic electron populations vary at different energy levels. Consider, for example, vinyl carbon atoms with charges varying at the σ and π levels, i.e., $\Delta q_i = \Delta q_{i\sigma} + \Delta q_{i\pi}$. In this case, the term $a_{ij}\Delta q_i$ becomes

$$a_{ij}\Delta q_i = a_{ij}^\sigma \Delta q_{i\sigma} + a_{ij}^\pi \Delta q_{i\pi} \tag{5.49}$$

indicating that a_{ij} represents now the appropriately weighted average of a_{ij}^σ and a_{ij}^π, i.e.,

$$a_{ij} = \frac{a_{ij}^\sigma \Delta q_{i\sigma} + a_{ij}^\pi \Delta q_{i\pi}}{\Delta q_{i\sigma} + \Delta q_{i\pi}}.$$

Taking advantage of the relationship $\Delta q_{i\sigma} = m\Delta q_{i\pi}$ (from Eq. 4.4), it appears that the appropriate average for vinyl carbons is simply

$$a_{ij} = \frac{ma_{ij}^\sigma + a_{ij}^\pi}{m + 1}. \tag{5.50}$$

The difficulty comes from the fact that calculated σ charges are considerably more basis set dependent than π charges, rendering a reliable theoretical evaluation of m illusory. Equation 5.50 is instrumental in numerical calculations of unsaturated hydrocarbons (Chapter 8). These applications suggest that $m \simeq -0.955$, a value which is consistent with the general conclusions (Chapter 4.7) drawn from the study of NMR shift–charge correlations.

These calculations wrap up the dull side of the a_{ij}'s. Let us now examine

TABLE 5.7. Calculation of a_{ij} (Eq. 5.47)

Atom	Population N_i (e)	$\partial E_i/\partial N_i$	$\partial^2 E_i/\partial N_i^2$	r_{ij} (Å)	a_{ij} (a.u.)
H $(1s)$	1.0117	−0.195		1.08 (CH)	$a_{HC} = -1.007$
C $(2s)$	1.44	−0.735	0.45	1.08 (CH)	$a_{CH} = -0.394$
				1.53 (CC)	$a_{CC} = -0.777$
				1.43 (CO)	$a_{CO} = -1.135$ (ether)
C $(2p)$	1.97	−0.200	0.37	1.08 (CH)	$a_{CH} = -0.276$ (carbonyl–C)
				1.53 (CC)	$a_{CC} = -0.659$ (carbonyl–C)
				1.22 (CO)	$a_{CO} = -1.182$ (carbonyl–C)
O $(2p)$	4.023	−0.316	0.50	1.22 (CO)	$a_{OC} = -1.065$ (carbonyl–O)[†]
	3.995	−0.331		1.43 (CO)	$a_{OC} = -0.804$ (ether–O)[‡]
C (σ)	3.00	−0.375		1.34 (C=C)	$a^{\sigma}_{C=C} = -0.802$ (sp^2-sp^2)
C (π)	1.00	−0.244		1.30 (C=C)	$a^{\pi}_{C=C} = -0.779$ (sp^2-sp^2)
C (σ)	3.00	−0.375		1.53 (CC)	$a^{\sigma}_{C_\alpha C} = -0.718$ (sp^2-sp^3)
C (π)	1.00	−0.244		1.55 (CC)	$a^{\pi}_{C_\alpha C} = -0.666$ (sp^2-sp^3)
C (σ)	3.00	−0.375		1.08 (CH)	$a^{\sigma}_{C_\alpha H} = -0.335$ (sp^2-H)
C (π)	1.00	−0.244		1.10 (CH)	$a^{\pi}_{C_\alpha H} = -0.2874$ (sp^2-H)

The selection of the appropriate $\partial E_i/\partial N_i$ derivatives presents no difficulties for sp^2 carbon and dialkylether oxygen atoms because *ab initio* population analyses clearly show the involvement of $2s$ electrons in the former case (Chapter 4.4) and of $2p$ electrons in the latter (Chapter 4.5). At carbonyl-C atoms, however, mainly $2p$ populations are affected by substitution. The appropriateness of this choice for $\partial E_i/\partial N_i$ is discussed in the text (Chapter 9.4). [†] Value for acetone; the a_{OC}'s for carbonyl groups must be recalculated for each compound of interest, using Eq. 5.47, because of the important contribution of the $(\partial^2 E_i/\partial N_i^2)\Delta q_i$ term. [‡] For the more negative oxygen atoms we find $a_{OC} = -0.800$ a.u. (Chapter 9.3). As regards the last six entries which concern the bonds formed by the sp^2 carbons of ethylene, see Chapter 8.2. Due to the inverse variations of σ and π charges, the final results must be calculated by means of Eq. 5.50 from the appropriate a^{σ}_{ij} and a^{π}_{ij} values indicated here: in this manner, it is found that $a_{C=C} = -0.291$, $a_{C_\alpha C} = 0.438$, and $a_{C_\alpha H} = 0.723$ a.u. (Chapter 8.4).

the more enlightening aspects and learn something about the charge-related effects determining how stable a molecule really is.

9. PHYSICAL INTERPRETATION OF THE a_{ij} PARAMETERS

The simple transfer of bonds from one molecule to another, described by the sum $\Sigma\varepsilon_{ij}^{\circ}$, implies transferring the bond-forming atoms, including their electron clouds, at the sacrifice of molecular electroneutrality. Actual molecules, of course, ensure their electroneutrality simply by allowing all atomic charges to assume their proper values, which are reflections of the appropriate molecular wave functions. The Δq_i's are the changes in net atomic charges satisfying this requirement. A negative Δq_i value corresponds to an actual increase of electron population at atom i. The a_{ij} coefficients, which measure the energetic effect of a unitary charge increment at atom

i, bonded to j, are necessarily negative (Eq. 5.47). Consequently, *any increase in electronic charge at the bond-forming atoms leads to a larger "bond energy" contribution, which is a stabilizing effect.**

The detailed features of these bond-stabilizing effects by electronic charges are most interesting. The effect of adding 0.001 electron (i.e., $\Delta q_i = -1$ me) is conveniently described by expressing the a_{ij}'s in kcal mol^{-1} me^{-1} units (1 a.u. = 627.51 kcal mol^{-1}). In this fashion, the a_{ij} parameters indicate that 1 me added to hydrogen stabilizes a CH bond by 0.632 kcal mol^{-1}, whereas 1 me added to carbon stabilizes a CC single bond by 0.488, a CH bond by 0.247, an ether-CO bond by 0.712, and a carbonyl-CO bond by 0.742 kcal mol^{-1}. Similarly, an electron enrichment of 1 me at an oxygen atom stabilizes an ether-CO bond by ~ 0.505 and a carbonyl-CO bond by ~ 0.668 kcal mol^{-1}, the precise values depending on the oxygen net charges.

These results answer a number of questions regarding charge-related energy effects. Consider, for example, a charge decrease of 1 me at a hydrogen atom in a carbonyl compound in favor of the oxygen atom, causing a loss in stability of 0.632 kcal mol^{-1} at the CH bond involved and a gain of 0.668 kcal mol^{-1} at the CO bond, for a net gain of 0.036 kcal mol^{-1}. This example shows that, *under circumstances which can be evaluated from the a_{ij}'s, a transfer of electronic charge accompanying isomerization can result in a relatively small change in total stability while the "location" of a certain portion of the overall stability is shifted from one part of the molecule to another*[34]. Of course, this type of reasoning applies only to comparisons involving molecules at equilibrium. With this understanding, the example just presented could apply to a comparison between, say, propanal and propanone, the latter compound being the more stable by ~ 6.1 kcal mol^{-1} in ΔE_a^* energy, which is largely, but not exclusively, traced back to an electron enrichment at the oxygen atom of the ketone at the expense of electronic charge at the hydrogen atoms.

For hydrocarbons, the simple rules deduced from the a_{ij} values describe by far the largest part of all the effects which govern their molecular stabilities and differentiate isomers from one another. Let us examine a few examples and consider, to begin with, an alkane C_β—C_α—H fragment. The transfer of 1 me from the hydrogen to the adjacent α-carbon destabilizes the CH bond by $0.632 - 0.247 = 0.385$ kcal mol^{-1} and stabilizes the C_α—C_β bond

---

*These conclusions do not always apply, however, with changes in net atomic charges expressed simply as *total* changes. Consideration should be given to the individual contributions (e.g., $2s$, $2p$, σ or π electrons) making up the total variations in charge, because these contributions may well combine to give resultant positive a_{ij}'s. A case in point is bond-formation involving charge variations at the π and σ levels, as with sp^2 carbons forming CH bonds ($a_{C\alpha H} = 0.723$ a.u.) or bonds with sp^3 carbon atoms ($a_{C\alpha C} = 0.438$ a.u.). Positive a_{ij}'s of this sort, resulting from a combination of effects described by Eq. 5.50, are not considered in the present discussion. The positive a_{ij}'s occurring in calculations of olefins are dealt with in Chapter 8.4. These reservations do not affect the conclusions presented in this Section, which are based on cases with negative a_{ij}'s.

by 0.488 kcal mol^{-1}. The net gain in stabilization is thus $0.488 - 0.385 = 0.103$ kcal mol^{-1} for this fragment. Had the transfer occurred to the β carbon, the C_α—H bond would have been destabilized by 0.632 and C_α—C_β stabilized by 0.488 kcal mol^{-1}. Moreover, additional C_β—H or C_β—C bonds would have been stabilized by 0.247, $viz.$, 0.488 kcal mol^{-1}. It follows therefrom that any electron enrichment on carbon atoms at the expense of electron populations at the hydrogen atoms results in a gain in molecular stability. This rule expresses, in a nutshell, the nature of the prime factors governing molecular stabilities of hydrocarbons. In comparisons between isomers or conformers, the more stable form has, on the whole, somewhat "weaker" CH bonds, but this is largely compensated by the stabilization of the carbon skeleton. So, for example, if we are asked why adamantane is more stable than twistane (by 9.1 kcal mol^{-1} in ΔE_a^* energy[35]), the answer is that the largest part by far of this energy difference is due to the fact that adamantane has less electrons on the H atoms and, consequently, more on the C atoms, than twistane. In a way, *the alkyl hydrogen atoms play the role of a reservoir of electronic charge which, under appropriate circumstances depending on molecular geometry, is called upon to stabilize bonds other than CH bonds, with a net gain in molecular stability.* This conclusion holds whenever the sum $\Sigma_j a_{ij}$, measuring the stabilization of all the bonds formed by atom i by an electronic charge added to it, is more negative than a_{HC}.

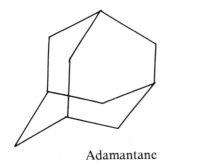

Adamantane

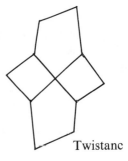

Twistane

With the present considerations, which will be expanded in due time with additional results, we have quite abruptly swung from a string of theoretical formulas to a numerical description of detailed aspects regarding the subtle fine-tuning of molecular energies in terms of electron distributions. Here we may rightfully wonder how accurate the above figures are. The only way of really knowing this is to carry out actual calculations using the appropriate charge distributions and to offer comparisons with experimental results. This is done in the following Chapters but, since this is not a mystery story, we may as well reveal the outcome here: for a body of ~ 100 molecules, the average deviation between calculated and experimental energies is 0.16 kcal mol^{-1}, which is well within experimental uncertainties. Before proceeding with numerical verifications, let us summarize briefly the various interlocking results described in this Chapter.

10. SUMMARY

The calculation of molecular energies is greatly facilitated by a separation of atomization energies into nonbonded and bonded contributions and by considering the latter as a sum of individual bond energy terms. Note that only one bond energy term is associated with any given bonded atom pair, meaning that a multiple bond (in conventional language) is also treated as *one* bond. In this perspective, the exact quantum mechanical formulation of atomic and molecular energies and the postulate that "chemical bonds" exist are combined to show that the portion of the total molecular atomization energy associated with a bond formed by atoms i and j (i.e., the "bond energy" ε_{ij}) can be expressed in terms of the derivatives $\partial \varepsilon_{ij}/\partial Z_i$ and $\partial \varepsilon_{ij}/\partial Z_j$, where Z_i and Z_j are the nuclear charges of atoms i and j (Eq. 5.35). From a set of $\partial \varepsilon_{ij}/\partial Z_i$ derivatives deduced for selected "reference" bonds, one obtains the corresponding "reference" bond contributions, ε_{ij}°. For example, taking ethane as the reference molecule for defining CC and CH bond energies, we find $\varepsilon_{CC}^\circ = 69.63$ and $\varepsilon_{CH}^\circ = 106.81$ kcal mol^{-1}.

Bond energies deduced for one molecule cannot be simply transferred to other molecules. The reason why a simple bond additivity scheme involving constant bond contributions cannot work is of fundamental importance. Indeed, the definition of constant bond energy terms implies an *a priori* selection of well-defined (constant) $\partial \varepsilon_{ij}/\partial Z_i$ derivatives (e.g., those of the ethane CC and CH bonds, to give the above ε_{CC}° and ε_{CH}° values). Now, these derivatives are assessed under the precise conditions that the internuclear distances and the electron distributions are kept constant. Hence, the transfer of constant bond energy terms from one molecule to another implies, ultimately, the construction of molecules using "atoms" having the same electron populations as in the molecule of reference. The important point is that this would not yield electroneutral molecules unless, of course, one denies any form of intramolecular charge transfer. For example, using the carbon and hydrogen "atoms" of ethane with net charges of 0.0351 and -0.0117 e, respectively, one obtains a methane "molecule" carrying an excess electronic charge of -0.0117 e. Consequently, before the failure of simple bond additivity schemes involving constant bond terms is blamed on "unidentified effects" of whatever nature, the first step to be made is to ensure electroneutrality of the molecules by restoring the appropriate electron distributions. From there on, the theory of the chemical bond becomes a theory of electron density.

The appropriate charge renormalization leads to an energy expression featuring the increments in net atomic charge, Δq_i and Δq_j, with respect to the charges of the atoms i and j in the reference bond. The a_{ij} and a_{ji}

parameters, which measure the energy effects due to unitary charge increments at atoms i and j, are derived from theory (Eq. 5.47) and numerical values are given. It is readily seen that any charge increase at the bond-forming atoms is a stabilizing effect and the outcome of charge redistributions accompanying isodesmic structural changes is now easy to evaluate.

"Steric effects", if interpreted as Coulomb-type interactions between nonbonded atoms, usually play only a negligible role in determining molecular stabilities. This is just as fine, because our analysis shows that the factors explaining the stereochemical effects are essentially contained in the description of the charge-dependent part of the bond energies, which is not surprising, considering that any stereochemical effect is reflected in the molecular wave function and, consequently, in the charge distributions.

REFERENCES

1. J.D. Cox and G. Pilcher, "Thermochemistry of Organic and Organometallic Compounds", Academic Press, New York, NY, 1970; W.C. Herndon, *Progress in Physical Organic Chemistry*, Vol. 9, Edited by A. Streitwieser Jr. and R.W. Taft, Wiley-Interscience, New York, NY (1972).
2. K.J. Laidler, *Can. J. Chem.*, **34**, 626 (1956); E.G. Lovering and K.J. Laidler, *Can. J. Chem.*, **38**, 2367 (1960); W.M. Tatevskii, V.A. Benderskii, and S.S. Yarovoi, "Rules and Methods for Calculating the Physico-Chemical Properties of Paraffinic Hydrocarbons", B.P. Mullins, Translation Ed., Pergamon Press, Oxford, 1961.
3. T.L. Allen, *J. Chem. Phys.*, **31**, 1039 (1959); A.J. Kalb, A.L.H. Chung, and T.L. Allen, *J. Am. Chem. Soc.*, **88**, 2938 (1966).
4. G.R. Somayajulu and B.J. Zwolinski, *Trans. Faraday Soc.*, **62**, 2327 (1966); G.R. Somayajulu and B.J. Zwolinski, *J. Chem. Soc., Faraday Trans. II*, **68**, 1971 (1972); *ibid.*, **70**, 967 (1974); J.B. Greenshields and F.D. Rossini, *J. Phys. Chem.*, **62**, 271 (1958).
5. B. Nelander and S. Sunner, *J. Chem. Phys.*, **44**, 2476 (1966).
6. S. Fliszár and J.-L. Cantara, *Can. J. Chem.*, **59**, 1381 (1981).
7. R.P. Feynman, *Phys. Rev.*, **56**, 340 (1939); T. Berlin, *J. Chem. Phys.*, **19**, 208 (1951).
8. S. Fraga, *Theor. Chim. Acta*, **2**, 406 (1964).
9. P. Politzer and R.G. Parr, *J. Chem. Phys.*, **61**, 4258 (1974).
10. P. Politzer and R.G. Parr, *J. Chem. Phys.*, **64**, 4634 (1976).
11. R.G. Parr, R.A. Donnelly, M. Levy, and W.E. Palke, *J. Chem. Phys.*, **68**, 3801 (1978).
12. P. Politzer, *J. Chem. Phys.*, **70**, 1067 (1979).
13. T. Anno, *J. Chem. Phys.*, **72**, 782 (1980).
14. S. Fliszár and M.-T. Béraldin, *J. Chem. Phys.*, **72**, 1013 (1980).
15. W.-P. Wang and R.G. Parr, *Phys. Rev. A*, **16**, 891 (1977).
16. S. Fliszár and H. Henry, *J. Chem. Phys.*, **67**, 2345 (1977).
17. S. Fliszár, *J. Chem. Phys.*, **69**, 237 (1978).
18. S. Fliszár and D. Salahub, *J. Chem. Phys.*, **69**, 3321 (1978).
19. R.J. Boyd, *J. Chem. Phys.*, **66**, 356 (1977).
20. R.G. Parr and S.R. Gadre, *J. Chem. Phys.*, **72**, 3639 (1980).

21. S. Fliszár, M. Foucrault, M.-T. Béraldin, and J. Bridet, *Can. J. Chem.*, **59**, 1074 (1981).
22. R.G. Parr, S.R. Gadre, and L.J. Bartolotti, *Proc. Nat. Acad. Sci. USA*, **76**, 2522 (1979).
23. K. Ruedenberg, *J. Chem. Phys.*, **66**, 375 (1977).
24. P. Politzer, *J. Chem. Phys.*, **64**, 4239 (1976).
25. A.B. Sannigrahi, B.R. De, and B. Guha Niyogi, *J. Chem. Phys.*, **68**, 784 (1978).
26. N.H. March, in "Theoretical Chemistry", Vol. 4, The Royal Society of Chemistry, Burlington House, London, 1981.
27. S. Fliszár, *J. Am. Chem. Soc.*, **102**, 6946 (1980).
28. G. Del Re, *Gazz. Chim. Ital.*, **102**, 929 (1972).
29. J.R. Platt, *Handb. Phys.*, **371c**, 188 (1961).
30. J.C. Slater, *Adv. Quantum Chem.*, **6**, 1 (1972); J.C. Slater, *Int. J. Quantum Chem., Symp.*, **3**, 727 (1970); J.C. Slater in "Computational Methods in Band Theory", P.M. Marcus, J.F. Janak, and A.R. Williams, Eds., Plenum Press, New York, London, 1971, p. 447; J.C. Slater, "The Self-Consistent Field for Molecules and Solids", Vol. 4, McGraw-Hill, Kuala-Lumpur, 1974.
31. G. Kean, M.Sc. Thesis, Université de Montréal, Montréal, 1974.
32. T.H. Dunning, *J. Chem. Phys.*, **53**, 2823 (1970).
33. K. Schwarz, *Phys. Rev. B*, **5**, 2466 (1972).
34. S. Fliszár and M.-T. Béraldin, *Can. J. Chem.*, **60**, 792 (1982).
35. J.K. Choi, M.J. Joncich, Y. Lambert, P. Deslongchamps, and S. Fliszár, *J. Mol. Struct. (Theochem.)*, **89**, 115 (1982).

CHAPTER 6

Energy Analysis of Saturated Hydrocarbons

1. INTRODUCTION

The purpose of the numerical calculations presented in this Chapter is twofold. In the first place they enable detailed comparisons with experimental energies. These comparisons provide an understanding of a number of major conformational effects in terms involving only basic physical effects (i.e., nuclear–electronic and nuclear–nuclear potential energies) and disclose trends regarding molecular stabilities at the vibrationless level, on the one hand, and vibrational energies on the other. Secondly, these numerical evaluations offer the opportunity to discuss the validity of the atomic charges, namely those of carbon atoms, which are used in energy calculations, and answer a few puzzling questions, e.g., why is the calculated bond contribution of the ethane CC bond, $\varepsilon_{CC}^{\circ} - 69.63$ kcal/mol, so different from the commonly accepted value,[1] ~ 80 kcal/mol?

The availability (or lack of it) of the required *complete* sets of experimental results is a factor which limits comparisons with theory. Indeed, while it is easy to derive ΔE_a^* atomization energies from Eq. 5.5

$$\Delta E_a^* = \sum_i n_i \left[\Delta H_f^{\circ}(A_i) - \frac{5}{2} RT \right] + \text{ZPE} + (H_T\text{-}H_0) - \Delta H_f^{\circ}$$

and, moreover, enthalpies of formation (ΔH_f°) as well as heat content energies ($H_T\text{-}H_0$) are known for a large collection of molecules, adequate information about zero-point energies (ZPE) is relatively scarce. On the theoretical side, the single most important difficulty rests with the accurate evaluation of atomic charges. In a number of cases, atomic charges can be deduced from appropriate, carefully established, correlations with nuclear magnetic resonance shifts (Chapter 4) so that, ultimately, we end up calculating molecular energies from NMR spectra. This type of approach, dictated by the intrinsic difficulties in obtaining reliable charges, obviously increases the number of molecules which can be investigated. On the other hand, it is also clear that, because of this way of proceeding, the molecules used in comparisons of theory with experiment belong to large classes of homologous compounds rather than covering a broad spectrum of chemically different situations. On the bright side, however, the definite loss in excitement is largely compensated by the regularity with which the present theoretical approach reproduces energy differences, even small ones, between closely related compounds. This Chapter is confined to the study of saturated hydrocarbons.

In the study of homologous series of molecules, like the C_nH_{2n+2} and C_nH_{2n} hydrocarbons, it is practical to express the energy formula for $\Delta E_a^{* \, bonds}$ (Eq. 5.48) by a general equation which holds for all the members of that series. Before doing so, however, it is instructive to consider bond-by-bond calculations of the individual ε_{ij}'s, using Eq. 5.46, as indicated in the following examples.

2. THE BOAT AND CHAIR FORMS OF CYCLOHEXANE

It is common knowledge that cyclohexane exists predominantly in its chair form. For the boat conformer, the direct observation of its NMR spectrum not being feasible, we resort to a calculation of the carbon–13 shifts using the methods and parameters described by Grant et al.[2], which gives δ 10.7 for carbons 1 and 4 and δ 16.5 (ppm from ethane) for the other four carbon atoms. It follows from Eq. 4.10, i.e., $\Delta q_C = -0.148\delta_C$ me relative to the ethane-C net charge, that $\Delta q_C = -1.584$ me for carbons 1 and 4 and $\Delta q_C = -2.442$ me for the other C atoms.

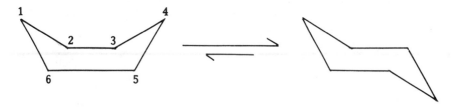

Using now Eq. 5.46, i.e., $\varepsilon_{ij} = \varepsilon_{ij}^{\circ} + a_{ij}\Delta q_i + a_{ji}\Delta q_j$, with $a_{CC} = -0.488$ kcal mol^{-1} me^{-1} (from Table 5.7), we find that each of the four CC bonds involving C–1 and C–4 is $0.488(1.584 + 2.442) = 1.965$ kcal mol^{-1} more stable than the ethane CC bond and that each of the remaining two CC bonds (i.e., C_2—C_3 and C_5—C_6) is more stable by 2.383 kcal mol^{-1}. The total gain in stability, relative to ethane CC bonds, is thus 12.63 kcal mol^{-1} for the carbon skeleton of boat cyclohexane.

Since the atomic charge of the ethane-C atom is taken at 35.10 me, it follows that the carbon net charges are $35.10 - 1.584 = 33.516$ me (C–1, –4) and 32.658 me (C–2, –3, –5, –6) for a total of 197.66 me on the carbon atoms and, consequently, -16.472 me (on the average) for each hydrogen atom. The hydrogen net charge in ethane is $q_H^\circ = -11.70$ me (from $q_C^\circ + 3q_H^\circ = 0$) and, therefore, $\Delta q_H = q_H - q_H^\circ = -4.772$ me. This average Δq_H value, while not giving a true reflection of the individual CH bonds, is sufficient for the correct evaluation of the total gain in stability of the CH part because each Δq_H enters the final sum with the same a_{HC} coefficient. Using the $a_{HC} = -0.632$ and $a_{CH} = -0.247$ kcal mol^{-1} me^{-1} values given in Table 5.7, Eq. 5.47 indicates that each of the four CH bonds formed by C–1 and C–4 is (in this "average" calculation) $(0.247 \times 1.584) + (0.632 \times 4.772) = 3.407$ kcal mol^{-1} more stable than the ethane CH bond, and that each of the other eight CH bonds has gained $(0.247 \times 2.442) + (0.632 \times 4.772) = 3.619$ kcal mol^{-1}, for a total gain in stability of 42.58 kcal mol^{-1} in the CH bonds.

Finally, the total stabilization of boat cyclohexane is 12.63 (CC bonds) + 42.58 kcal mol^{-1} (CH bonds) with respect to a hypothetical cyclohexane molecule constructed from ethane CC and CH bonds whose contributions are $\varepsilon_{CC}^\circ = 69.63$ and $\varepsilon_{CH}^\circ = 106.81$ kcal mol^{-1} (Table 5.6). The total stabilization, 55.21 kcal mol^{-1}, is the $\Sigma_i \Sigma_j a_{ij} \Delta q_i$ term of Eq. 5.48, and $\Delta E_a^{* \, bonds}$ is 1754.71 kcal mol^{-1}.

Similarly, for the chair form of cyclohexane, we obtain from δ 21.8 (ppm from ethane) that $\Delta q_C = -3.227$ me, giving a stabilization of $2 \times 0.488 \times 3.227 = 3.150$ kcal mol^{-1} for each CC bond. Since the average charge on the hydrogen atoms is now -15.937 me and, thus, $\Delta q_H = -4.237$ me, each CH bond is stabilized by $(0.247 \times 3.227) + (0.632 \times 4.237) = 3.475$ kcal mol^{-1}. The total gain in stability is, hence, 18.90 (CC bonds) + 41.70 kcal mol^{-1} (CH bonds) relative to the ethane bonds, giving $\Sigma_i \Sigma_j a_{ij} \Delta q_i = 60.60$ and $\Delta E_a^{* \, bonds} = 1760.10$ kcal mol^{-1}.

Consequently, comparing now the boat and chair forms, it is deduced that the 12 CH bonds are more stable in the boat conformer by 0.88 kcal mol^{-1} and that the carbon skeleton of chair cyclohexane is more stable than that of the boat form by 6.27 kcal mol^{-1}, giving a total difference in stability of 5.39 kcal mol^{-1} favoring the chair conformer. This theoretical

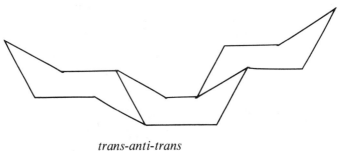

trans-anti-trans

$\Delta H_f^\circ = -52.74$ kcal mol^{-1}

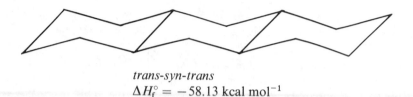

trans-syn-trans
$\Delta H_f^\circ = -58.13$ kcal mol^{-1}

result agrees with the measured energy difference (5.39 kcal mol^{-1}) between the *trans-anti-trans* and *trans-syn-trans*-perhydroanthracenes[3], which differ only because of the center boat in the former compound.

So far, the energy calculations for the two forms of cyclohexane were done with reference to molecules in their hypothetical vibrationless state. On the other hand, force-field calculations of the fundamental frequencies of boat and chair cyclohexane[4] indicate that, within the precision of this type of analysis, the two forms differ by no more than ± 0.1 kcal mol^{-1} from one another in total vibrational energy (Eqs. 5.6 and 5.7) so that, ultimately, the difference in *enthalpy* of formation between the two conformers can be reasonably estimated at ~ 5.4 (± 0.1) kcal mol^{-1}. Consequently, using the experimental value $\Delta H_f^\circ(\text{chair}) = -29.5$, it follows that (in theory) $\Delta H_f^\circ(\text{boat}) = -24.1$ kcal mol^{-1} (gas, 298.15 K).

"Bond-by-bond" calculations like those presented here for illustrative purposes are lenghty. If one is interested only in the final energy result, it is advantageous to derive and use general formulas. This is done in the next section, after which we can tackle additional conformational problems in a more straightforward way and offer direct comparisons with experimental results.

3. GENERAL FORMULAS FOR SATURATED HYDROCARBONS

Saturated hydrocarbons $C_nH_{2n+2-2m}$ containing m ($\geqslant 0$) six-membered rings are not only the backbone of organic chemistry but, fortunately, also represent the largest class of compounds for which we possess sufficient experimental results, namely, the thermochemical and spectroscopic data required for deducing the ΔE_a^*'s, and comprehensive information about atomic charges.

To begin with the calculation of the $\Sigma \varepsilon_{ij}^\circ$ and $\Sigma_i \Sigma_j a_{ij} \Delta q_i$ terms of Eq. 5.48, we note that the number of CC bonds in the molecule is $n - 1 + m$ and that the number of CH bonds is $2n + 2 - 2m$. Consequently, it follows that

$$\sum \varepsilon_{ij}^\circ = (n - 1 + m)\varepsilon_{CC}^\circ + (2n + 2 - 2m)\varepsilon_{CH}^\circ \qquad (6.1)$$

Next, we observe that the $a_{CC}\Delta q_C$ term occurs once for each CC bond formed by the C atom to which Δq_C refers, so that its total contribution to carbon-carbon bonds is $a_{CC}N_{CC}\Delta q_C$, where N_{CC} is the number of CC bonds this C atom is engaged in. Similarly, each $a_{CH}\Delta q_C$ term occurs $4 - N_{CC}$ times (i.e., once for each CH bond) and, finally, each $a_{HC}\Delta q_H$ term only once. The sum over the entire molecule is then

$$\sum_i \sum_j a_{ij}\Delta q_i = (a_{CC} - a_{CH})\sum N_{CC}\Delta q_C + 4a_{CH}\sum \Delta q_C + a_{HC}\sum \Delta q_H. \quad (6.2)$$

The sum $\Sigma \Delta q_H = \Sigma(q_H - q_H^\circ)$ is obtained from charge normalization, i.e., $\Sigma q_H = -\Sigma q_C$. It follows that $\Sigma \Delta q_H = -\Sigma(q_C - q_C^\circ) - nq_C^\circ - (2n + 2 - 2m)q_H^\circ$ and, because $q_C^\circ + 3q_H^\circ = 0$ (for ethane), that

$$\sum \Delta q_H = -\sum \Delta q_C + (n - 2 + 2m)q_H^\circ$$

The final expression is now derived from Eq. 6.2, which becomes

$$\sum_i \sum_j a_{ij}\Delta q_i = (a_{CC} - a_{CH})\sum N_{CC}\Delta q_C + (4a_{CH} - a_{HC})\sum \Delta q_C$$
$$+ (n - 2 + 2m)a_{HC}q_H^\circ.$$

The $(a_{CC} - a_{CH})$ and $(4a_{CH} - a_{HC})$ terms occur in every calculation for molecules containing alkyl groups. Defining now

$$\lambda_1 = a_{CC} - a_{CH} \qquad (6.3)$$

$$\lambda_2 = 4a_{CH} - a_{HC} \qquad (6.4)$$

we obtain

$$\sum_i \sum_j a_{ij}\Delta q_i = \lambda_1 \sum N_{CC}\Delta q_C + \lambda_2 \sum \Delta q_C + (n - 2 + 2m)a_{HC}q_H^\circ \quad (6.5)$$

which is, together with Eq. 6.1, the basic working formula in actual calculations of $\Delta E_a^{* \,bonds}$. Alternate convenient expressions for $\Delta E_a^{* \,bonds}$ are obtained by observing that $\Delta E_a^{* \,bonds}(C_2) = \varepsilon_{CC}^\circ + 6\varepsilon_{CH}^\circ$ is the value for ethane, and that for methane we have $\Delta E_a^{* \,bonds}(C_1) = 4\varepsilon_{CH}^\circ + \lambda_2 \Delta q_C(C_1) - a_{HC}q_H^\circ$ (from Eqs. 6.1 and 6.5), where $\Delta q_C(C_1) = q_C(\text{methane}) - q_C^\circ$. In this manner we deduce from Eqs. 6.1 and 6.5 that[5]

$$\Delta E_a^{* \,bonds} = (1 - m)\Delta E_a^{* \,bonds}(C_2)$$
$$+ (n - 2 + 2m)[\Delta E_a^{* \,bonds}(C_2) - \Delta E_a^{* \,bonds}(C_1)] \quad (6.6)$$
$$+ \lambda_1 \sum N_{CC}\Delta q_C + \lambda_2[(n - 2 + 2m)\Delta q_C(C_1) + \sum \Delta q_C].$$

Finally, taking advantage of the identity

$$(n - 2 + 2m)q_H^\circ = -\frac{1}{2}(n - 1 + m)q_C^\circ - \frac{1}{4}(2n + 2 - 2m)q_H^\circ$$

it also follows, in the same fashion, that[5]

$$\Delta E_a^{* \,bonds} = (n - 1 + m)\left(\varepsilon_{CC}^\circ - \frac{a_{HC}q_C^\circ}{2}\right)$$
$$+ (2n + 2 - 2m)\left(\varepsilon_{CH}^\circ - \frac{a_{HC}q_H^\circ}{4}\right) \quad (6.7)$$
$$+ \lambda_1 \sum N_{CC}\Delta q_C + \lambda_2 \sum \Delta q_C.$$

Applications of Eqs. 6.6 and 6.7 are presented further below, in the discussion of approximations facilitating numerical calculations of atomization energies. The success of the entire operation rests now with the validity of the charge analyses giving the Δq_i's.

4. DEFINITION OF CHARGES SATISFYING EQS. 5.46–5.48

Two aspects must be covered in the definition of carbon atomic charges, namely, their correct relative ordering and their absolute values. The forthcoming discussion is restricted to normal and branched saturated hydrocarbons and to six-membered cycloalkanes. Other cases will be considered in order of appearance. The hydrogen net charges represent no special problem in the present case, as they are derived from charge normalization.

The correct relative ordering of the charges is, of course, a necessary condition in any study of property–charge relationships (Chapter 3.1). It is determined by the parameter p appearing in the useful approximations applicable to saturated hydrocarbons, i.e.,

$$q_H = q_H^{\text{Mulliken}} - p$$

$$q_C = q_C^{\text{Mulliken}} + N_{CH} p$$

where q_H^{Mulliken} and q_C^{Mulliken} are Mulliken's charges, N_{CH} is the number of H atoms bonded to C, and p is the departure from the usual halving of the C—H overlap population, for one CH bond. The appropriate p for the problem at hand was determined as follows. Applying Eq. 5.46, in which $\Delta q_C = q_C - q_C(\text{ethane})$ and $\Delta q_H = q_H - q_H(\text{ethane})$, we have expressed q_C and q_H as indicated above by using Mulliken populations as input and leaving p as the unknown to be determined. The Mulliken charges were derived from fully optimized STO–3G calculations (Table 2.8). The $\Delta E_a^{*\text{bonds}}$'s constructed in this fashion (Eq. 6.7) were compared with their experimental counterparts, and p was determined by least-square analysis. This procedure amounts to an experimental partitioning of overlap populations. For fully optimized STO–3G charges, we obtain[5] $p = (30.3 \pm 0.3) \times 10^{-3}$ e. On the other hand, the same set of Mulliken charges, when compared to the corresponding ^{13}C NMR shifts, yields $p = 30.12 \times 10^{-3}$ e and gives Eq. 4.9, i.e., $\delta_C = -237.1\Delta q_C/q_C^\circ$. Consequently, since the same definition of charge satisfies the appropriate equations for $\Delta E_a^{*\text{bonds}}$ and for δ_C, we can use the latter (Eq. 4.9) for deriving the required Δq_C's. Detailed studies (Chapters 2, 3) have shown that this sort of analysis holds independently of the LCAO–MO method selected for calculating Mulliken charges. However, while the ordering of the carbon net charges (i.e., q_C/q_C°) is uniquely defined, we are presently unable to derive theoretically the reference net charge q_C° in a satisfactory manner. The q_C°'s corresponding to STO–3G, $7s3p|3s$, and 6–31G calculations, for example, are 0.0694, 0.060, and 0.058 e, respectively, while the value resulting from the forthcoming numerical analysis is 0.0351 e.

The appropriate q_C° value can be evaluated[5] by rearranging Eqs. 6.6 and 6.7 to give

$$2n\varepsilon_{CC}^\circ = \Delta E_a^{*\text{bonds}} + n[\Delta E_a^{*\text{bonds}}(C_2) - 2\Delta E_a^{*\text{bonds}}(C_1) + 2\lambda_2\delta_C(C_1)$$
$$+ a_{HC}q_C^\circ] + (1 - m)[\Delta E_a^{*\text{bonds}}(C_2) - 2\Delta E_a^{*\text{bonds}}(C_1)$$
$$+ 2\lambda_2\delta_C(C_1)] - \lambda_1\sum N_{CC}\delta_C - \lambda_2\sum\delta_C$$

and by taking advantage of the fact that ε_{CC}° is constant by definition. In this equation, the Δq_C's are now expressed in terms of chemical shifts relative to ethane, i.e., $\Delta q_C = -(q_C^\circ/237.1)\delta_C$ (from Eq. 4.9), which ensures the proper ordering of the carbon net charges. Consequently, using the theoretical a_{ij}'s (Table 5.7), we write (in kcal mol^{-1} ppm^{-1} units) $\lambda_1 = 0.383(627.51/237.1)q_C^\circ$ and $\lambda_2 = 0.569(627.51/237.1)q_C^\circ$ and find that for $q_C^\circ \simeq 0.035$ e, ε_{CC}° remains constant at ~ 69.7 kcal mol^{-1} within the limits set by experimental uncertainties. Our optimum choice, $q_C^\circ = 0.0351$ e, yields $\lambda_1 = 0.0356$ and $\lambda_2 = 0.0529$ kcal mol^{-1} ppm^{-1} and $\varepsilon_{CC}^\circ = 69.63$ kcal mol^{-1}. While this choice remains open to discussion regarding its precision, we feel presently unable to go beyond the present level of sophistication because of the uncertainties affecting experimental results and the evaluation of precise nonbonded interactions using Eq. 5.30. At this level, the calculations are well within experimental uncertainties.

A noteworthy point can be made using the empirical λ_1 and λ_2 values indicated above. Indeed, a valid test for the theory is offered by the comparison of the empirical $\lambda_2/\lambda_1 = 1.486$ ratio, deduced from selected experimental ΔE_a^* results[5], with its theoretical counterpart

$$\frac{4a_{CH} - a_{HC}}{a_{CC} - a_{CH}} = \frac{(\partial E_C/\partial N_C) - (6/7)(\partial E_H/\partial N_H)}{(-3/7)(Z_C^{\text{eff}}\langle r_{CC}^{-1}\rangle^\circ - Z_H^{\text{eff}}\langle r_{CH}^{-1}\rangle^\circ)} = 1.483$$

which is calculated from the results given in Table 5.7. The point is that the empirical ratio, assuming estimates of nonbonded interactions for q_C° in the neighborhood of 35×10^{-3} e, is nearly independent of q_C°, and so is the theoretical ratio in which only the derivatives are slightly affected by the particular choice for q_C°.

The results discussed in this Section, i.e., $q_C^\circ = 35.1$ me and $\Delta q_C = -(q_C^\circ/237.1)\delta_C = -0.148\delta_C$ complete the information which is required for numerical applications of Eqs. 5.46–5.48 to saturated hydrocarbons. We can now examine individual molecules.

5. NUMERICAL APPLICATIONS: SATURATED HYDROCARBONS

The discussion of energy formulas has, so far, centered on charge distributions and on their role in local energy effects. With the relationship $\Delta q_C = -0.148\delta_C$ me describing carbon charges in the series of molecules under study it is now more convenient to express energies in terms of ^{13}C chemical shifts (relative to ethane). The appropriate theoretical parameters are then $\lambda_1 = 0.0356$, $\lambda_2 = 0.0529$ kcal mol^{-1} ppm^{-1}, and $a_{HC}q_H^\circ = 7.393$ kcal mol^{-1}. In this way, Eq. 6.5 becomes (in kcal mol^{-1})

$$\sum_i \sum_j a_{ij}\Delta q_i = 0.0356\sum N_{CC}\delta_C + 0.0529\sum\delta_C + 7.393(n - 2 + 2m). \quad (6.8)$$

This equation certainly represents the most direct way of obtaining the desired results. For example, from the NMR shifts described for boat cyclohexane (Sect. 2) it follows that $\Sigma N_{CC}\delta_C = 174.8$ and $\Sigma\delta_C = 87.4$ ppm and, therefrom, $\Sigma_i\Sigma_j a_{ij}\Delta q_i = 55.20$ kcal mol^{-1}. Similarly, we find for the chair form that $\Sigma N_{CC}\delta_C = 261.6$ and $\Sigma\delta_C = 130.8$ ppm, giving $\Sigma_i\Sigma_j a_{ij}\Delta q_i = 60.59$ kcal mol^{-1}. The energy difference of 5.39 kcal mol^{-1} favoring the chair conformer is easily traced back to the increase in electronic charge at its carbon atoms, which is reflected in the larger (downfield) $\Sigma\delta_C$ value.

This example illustrates an observation which, in fact, holds generally true. It can be expressed as a rule of thumb[6], dictated by Eq. 6.8. Since ^{13}C shifts are farther downfield as sp^3 carbon atoms are closer to electroneutrality (i.e., less positive) and both λ_1 and λ_2 are positive quantities (in kcal mol^{-1} ppm^{-1} units), it follows that "*in comparisons between isomers or conformers, the more stable compound is that whose carbon skeleton best approaches electroneutrality, which is reflected by larger (downfield) δ_C values*". Since any gain in electronic charge at the carbon atoms is made at the expense of hydrogen charges (when going from one isomer or conformer to another) we find here an echo of a conclusion reached earlier (Chapter 5.9) from an inspection of the a_{ij} coefficients, namely that the alkyl hydrogen atoms play the role of a reservoir of electronic charge which, under circumstances depending on molecular geometry, is called upon to stabilize bonds other than CH bonds, with a net gain in molecular stability.

There is a point, however, which does not become immediately clear, just from an inspection of Eq. 6.8, i.e., the physical origin of the relatively large $\Sigma_i\Sigma_j a_{ij}\Delta q_i$ contributions. This aspect of the question is more readily understood by examining the detailed "bond-by-bond" analysis presented in Sect. 2. It has appeared that in the chair and boat cyclohexanes both the carbon *and* the hydrogen atoms are electron-richer than in ethane. This is, of course, due to the fact that in cyclohexane only two H atoms withdraw electronic charge from the carbon atom to which they are attached (leaving a little more electronic charge behind, as compared with the ethane-C atom) and, moreover, that only two instead of three hydrogen atoms share the total charge which is missing on the carbon atom. In fact, had cyclohexane been constructed using carbon and hydrogen "atoms" like those of ethane, with net charges $q_C^\circ = 35.1$ and $q_H^\circ = -11.7$ me, then the cyclohexane "molecule" would have been electron deficient by 70.2 me. Generally speaking, a large part of the $\Sigma_i\Sigma_j a_{ij}\Delta q_i$ energy is simply accounted for by the required charge renormalization, while the fine-tuning which differentiates isomers or conformers from one another is found in the $\lambda_1\Sigma N_{CC}\delta_C + \lambda_2\Sigma\delta_C$ term.

Clearly, here we meet again with the basic reason at the origin of the non-transferability of bond energy terms (Chapter 5.6).

This answer triggers now another question, related to well-known facts: why are simple bond additivity schemes so successful in their own right, provided they account in one way or another for relatively "minor" (from the energy standpoint) departures from exact additivity (e.g., in terms of "steric effects"), although charge effects are not explicitly taken into consideration? This question is best discussed with the help of Eq. 6.7 which, indeed, reduces formally to a simple bond additivity scheme, respecting electroneutrality, provided all Δq_C's are set equal to 0. In this manner we achieve a token charge renormalization by assuming constant atomic charges at the carbon atoms and letting the hydrogen atoms pick up whatever is necessary to maintain electroneutrality. The "apparent" CC and CH bond contributions, to be multiplied by the appropriate numbers of CC and CH bonds in the molecule, now become[5]

$$\varepsilon^\circ_{CC} - \frac{a_{HC}q^\circ_C}{2} \simeq 80.7 \text{ kcal mol}^{-1},$$

$$\varepsilon^\circ_{CH} - \frac{a_{HC}q^\circ_H}{4} \simeq 105 \text{ kcal mol}^{-1}.$$

Clearly, these quantities are not the true CC and CH bond energies, because of the involvement of the $a_{HC}q^\circ_C/2$ and $a_{HC}q^\circ_H/4$ terms arising from charge normalization. Incidentally, the above ~ 80.7 and ~ 105 kcal mol^{-1} figures closely resemble those deduced in empirical fashion by Allen[7] (82.31 and 104.73 kcal mol^{-1}, for ethane). This aspect of the theory concerning "apparent" CC and CH contributions of ~ 80.7 and ~ 105 kcal mol^{-1}, respectively, is worth mentioning because it explains the origin of this sort of numerical values from empirical correlations and the fallacy underlying them. It is also clear, however, that with the use of "apparent" bond energies the gap between calculated and observed energies is considerably narrowed and, thus, certainly easier to account for by means of appropriate empirical adjustments.

This concludes the description of our working formula, Eq. 6.8, and the discussion of a few related topics. We may now proceed with applications intended to illustrate its use and consider, to begin with, selected problems in conformational analysis.

The first example refers to the conformational equilibrium between the axial and equatorial forms of methylcyclohexane. The low-temperature ^{13}C shifts of the C–3, C–5, and methyl carbon atoms of axial methylcyclohexane are ~ 6 ppm upfield from those of the equatorial form[8], thus decreasing $\Sigma N_{CC}\delta_C$ and $\Sigma\delta_C$ by 30 and 18 ppm, respectively. Following Eq. 6.8, the energy of the axial form is then $30\lambda_1 + 18\lambda_2 = 2.02$ kcal mol^{-1} lower than that of the equatorial conformer (in terms of $\Delta E_a^{*\text{bonds}}$) meaning that the latter is more stable than the axial form. We can now estimate the relative energies of the various forms of butane. Taking the boat and chair forms of cyclohexane as models, we assume a similar pattern in ^{13}C shifts for eclipsed and *gauche* butane as in the model compounds (see Sect. 2), i.e., a decrease of 11.1 and 5.3 ppm for the terminal and central C atoms of eclipsed butane. The latter is thus $43.4\lambda_1 + 32.8\lambda_2 = 3.28$ kcal mol^{-1} less stable than *gauche* butane. Similarly, taking axial and equatorial methylcyclohexane as models for the *gauche* and *anti* forms of butane, the ~ 6 ppm upfield shift of the C–3, C–5, and CH$_3$ carbons observed for the axial conformer results in a destabilization of $12\lambda_1 + 12\lambda_2 = 1.06$ kcal mol^{-1} of *gauche* butane with respect to its *anti* form. These conformational results[6], which are entirely derived from ^{13}C shifts using Eq. 6.8, are in general agreement with currently accepted values, moreover as for these structures the variations in zero-point and thermal energies represent, in all likelihood, only minor contributions[9].

These examples were discussed referring only to the changes in the $\lambda_1 \Sigma N_{CC}\delta + \lambda_2 \Sigma \delta_C$ part of $\Sigma_i \Sigma_j a_{ij} \Delta q_i$, describing adequately the changes in ΔE_a^{*bonds} energy accompanying isodesmic structural changes. Using now Eq. 6.8 in its entirety, and calculating also the corresponding $\Sigma \varepsilon_{ij}^{\circ}$ sum, we obtain ΔE_a^{*bonds} (Eq. 5.48) and, therefrom, $\Delta E_a^* = \Delta E_a^{*bonds} - E_{nb}^*$ (Eqs. 5.29 and 5.30). The calculated ΔE_a^*'s are compared in Table 6.1 with their experimental counterparts, deduced from Eq. 5.5. The agreement could hardly be any better, the average deviation (0.16 kcal mol^{-1}) being well within experimental uncertainties[5].

Although the examples are still limited in number, mainly because of the scarcity of adequate ZPE + H_T-H_0 data, it is already clear that the role of Coulomb-type interactions between nonbonded atoms is only a very minor one. The charge-dependent $\Sigma_i \Sigma_j a_{ij} \Delta q_i$ part associated with bonded interactions is by far the leading term accounting for the energetic differences between isomers or conformers. This observation is, in fact, quite a general one: it also holds true in other classes of compounds (ethers, carbonyl compounds, etc.).

Additional examples verifying the master equations (5.46–5.48) in the hydrocarbon series are presented in Chapter 7, after a study dealing explicitly with the role of vibrational contributions to total molecular energies. At this stage, however, it appears more convenient to study an approximation for ΔE_a^* which does not require detailed and lenghty calculations of the small nonbonded energies. Let us first examine these energies a little more in detail.

6. NONBONDED INTERACTION ENERGIES

The whole idea behind this study of nonbonded interaction energies is to get rid of them elegantly, at least as explicit terms requiring separate calculations. In order to achieve this, we must know how they evolve with increasing molecular size and how they depend on structural features. In short, we shall examine to what extent nonbonded Coulomb-type interactions are "addi-

TABLE 6.1. Comparison between Calculated and Experimental Atomization Energies of Saturated Hydrocarbons (kcal/mol)

Molecule	ΔH_f°(298.15, gas)	ZPE	$H_T - H_0$	$\Sigma \varepsilon_{ij}^\circ$	$\Sigma_i \Sigma_j a_{ij} \Delta q_i$	E_{nb}^*	ΔE_a^* calcd.	ΔE_a^* exptl.
Methane	-17.89	27.1	2.40	427.22	-7.82	0.09	419.31	419.24
Ethane	-20.04	45.16	2.86	710.47	0.00	-0.07	710.54	710.54
Propane	-25.02	62.43	3.51	993.71	10.38	-0.20	1004.29	1004.07
Butane	-30.03	79.73	4.65	1276.96	20.82	-0.32	1298.10	1298.15
Isobutane	-32.07	79.61	4.28	1276.96	22.78	-0.28	1300.02	1299.70
Pentane	-35.00	97.20	5.63	1560.20	31.25	-0.44	1591.90	1592.18
Isopentane	-36.92	96.84	5.30	1560.20	32.58	-0.39	1593.17	1593.43
Neopentane	-40.27	96.27	5.03	1560.20	35.54	-0.32	1596.06	1595.94
Hexane	-39.96	114.37	6.62	1843.45	41.69	-0.55	1885.69	1885.95
2-Methylpentane	-41.66	114.1	6.10	1843.45	43.05	-0.50	1887.00	1886.86
3-Methylpentane	-41.02	114.1	6.15	1843.45	42.38	-0.50	1886.33	1886.27
2,2-Dimethylbutane	-44.35	113.60	5.91	1843.45	44.53	-0.43	1888.41	1888.86
2,3-Dimethylbutane	-42.49	113.73	5.92	1843.45	43.32	-0.39	1887.16	1887.14
2,2,3-Trimethylbutane	-48.96	130.61	6.70	2126.69	55.00	-0.46	2182.15	2181.90
Cyclohexane	-29.50	103.30	4.24	1699.47	60.60	-0.73	1760.80	1760.82
Methylcyclohexane	-36.99	120.50	5.23	1982.72	73.19	-0.83	2056.74	2057.13
trans-Decalin	-43.52	160.68	6.22	2688.47	125.83	-1.26	2815.56	2815.50
Adamantane	-30.65	148.49	5.05	2544.49	142.18	-1.31	2687.98	2688.05
Bicyclo[2.2.2]octane	-23.75	125.89	4.94	2121.98	95.50	-1.03	2218.51	2218.40

The $\Sigma \varepsilon_{ij}^\circ$ sums were calculated using $\varepsilon_{CC}^\circ = 69.633$ and $\varepsilon_{CH}^\circ = 106.806$ kcal/mol, in agreement with the results given in: S. Fliszár, J. Am. Chem. Soc., **102**, 6946 (1980). The sources of the ΔH_f°, ZPE and $H_T - H_0$ values used for deriving the experimental ΔE_a^*'s (Eq. 5.5), as well as the appropriate $\Sigma N_{CC} \delta$ and $\Sigma \delta$ results for use in calculations of the $\Sigma_i \Sigma_j a_{ij} \Delta q_i$ terms (Eq. 6.5, viz. 6.8), are indicated in: H. Henry, G. Kean, and S. Fliszár, J. Am. Chem. Soc., **99**, 5889 (1977). The E_{nb}^* results are extracted from the reference cited above.

tive", at least in an approximate manner. This endeavor, of course, calls for a definition of "additivity". Its formulation is presented here for the $C_nH_{2n+2-2m}$ hydrocarbons[10].

Let X be a molecular property (e.g., E_{nb}^*) and $X(2)$, $X(1)$ the corresponding values for ethane and methane, respectively. If X is an exactly additive property, then

$$X = (1 - m)X(2) + (n - 2 + 2m)[X(2) - X(1)] \tag{6.9}$$

where $X(2) - X(1)$ is the change in X on going from methane to ethane, i.e., the contribution of one CH_2 group. The meaning of Eq. 6.9 is obvious for noncyclic compounds ($m = 0$). For example, the X value for propane is that of ethane plus the increment corresponding to one added CH_2 group. For cyclohexane ($m = 1$) which consists of $n - 2 + 2m = 6$ CH_2 groups, the $(1 - m)X(2)$ term of Eq. 6.9 cancels. Decalin is constructed from two cyclohexane units. In this case $n - 2 + 2m$ accounts for 12 CH_2 groups, but one additional ethane $X(2)$ contribution (i.e., that of two CH_2 and two H atoms) is subtracted with respect to cyclohexane, i.e., a total of two $X(2)$ contribu-

TABLE 6.2. Nonbonded Coulomb Interaction Energies E_{nb}^* (Eq. 5.30)

Molecule	E_{nb}^* (kcal/mol)
1 Methane	0.09
2 Ethane	−0.07
3 Propane	−0.20
4 Butane	−0.32
5 Isobutane	−0.28
6 Pentane	−0.44
7 Isopentane	−0.39
8 Neopentane	−0.32
9 2,2-Dimethylbutane	−0.43
10 2,3-Dimethylbutane	−0.39[†]
11 2,2,3-Trimethylbutane	−0.46
12 2,2,3,3-Tetramethylbutane	−0.44
13 Cyclohexane	−0.73
14 Bicyclo[2.2.2]octane	−1.03
15 Bicyclo[3.3.1]nonane	−1.13
16 *trans*-Decalin	−1.26
17 *cis*-Decalin	−1.23
18 Adamantane	−1.31
19 Iceane	−1.61

[†] Calculated for the statistical average of one *anti* (−0.40) and two *gauche* (−0.39) forms, as discussed in: L. Lunazzi, D. Macciantelli, F. Bernardi, and K.U. Ingold, *J. Am. Chem. Soc.*, **99**, 4573 (1977).

tions with respect to noncyclic alkanes. Similar arguments applied to other polycyclic saturated hydrocarbons verify the validity of Eq. 6.9 as a test for exact additivity.

Total Coulomb interactions between nonbonded centers were calculated from Eq. 5.30, using atomic charges corresponding to the scale defined by $q_C^\circ = 35.1$ me. It must be borne in mind, however, that this is a rough approximation, whose merits are difficult to assess. Numerical analyses were made assuming ~ 4 times larger nonbonded interactions than those given by Eq. 5.30 or, at the other extreme, assuming that $E_{nb}^* = 0$ in all cases (see below). It turns out that the overall results are not too severely affected by the precise definition of E_{nb}^*, although it would appear that the order of magnitude predicted in the present estimates consistently improves the overall agreement between calculated and experimental atomization energies. With these reservations in mind, it appears reasonable to use the estimates given by Eq. 5.30. The results are indicated in Table 6.2.

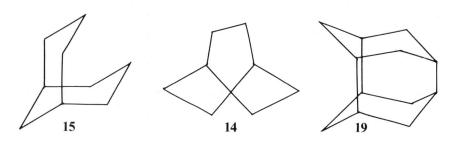

Attractive (i.e., stabilizing) interactions are negative. Of course, it is not surprising that branching increases repulsive (destabilizing) contributions, but the effect does not appear to be large in most cases. Let us now examine the situation which would result from a neglect of nonbonded Coulomb energy differences between isomers. In this approximation, we ask whether Coulomb interactions behave, at least to some extent, as if they were additive. This is simply done with the aid of Eq. 6.9 which expresses exact additivity. This equation, applied to the energies given in Table 6.2, yields the result presented in Fig. 6.1.

This result is self-explanatory. Nonbonded Coulomb interactions behave in general in a "quasi-additive" manner in terms of Eq. 6.9 if we agree upon accepting an uncertainty of ~ 0.1 kcal mol^{-1} due to the neglect of differences between isomers. Branching causes a systematic trend toward higher energies (repulsive destabilization) but situations of extreme steric crowding are required, such as those encountered in 2,2,3,3-tetramethylbutane and, to a lesser extent, in 2,2,3-trimethylbutane, in order to produce sizeable departures from "quasi-additivity", i.e., from Eq. 6.9. With these reservations in mind, we examine now the way of taking advantage of the "quasi-additivity" of nonbonded interactions in saturated hydrocarbons.

FIGURE 6.1. Additivity test for Coulomb nonbonded interaction energies by means of Eq. 6.9. The radii of the circles represent an uncertainty of ~ 0.03 kcal mol^{-1}. The numbering corresponds to the results indicated in Table 6.2.

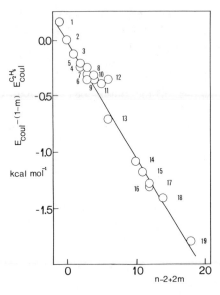

7. APPROXIMATE ΔE_a^* CALCULATIONS OF SATURATED HYDROCARBONS

First of all, following Eq. 6.9, we write the equation

$$E_{nb}^* \simeq (1 - m)E_{nb}^*(C_2) + (n - 2 + 2m)[E_{nb}^*(C_2) - E_{nb}^*(C_1)]$$

which expresses the "quasi-additive" behavior of nonbonded interactions. Next, we compare this result with Eq. 6.6 and, using also Eq. 5.29, write $\Delta E_a^{*bonds} - E_{nb}^*$ (for the molecule under study), $\Delta E_a^*(C_2) = \Delta E_a^{*bonds}(C_2) - E_{nb}^*(C_2)$ (for ethane), and $\Delta E_a^*(C_1) = \Delta E_a^{*bonds}(C_1) - E_{nb}^*(C_1)$ (for methane). In this fashion we obtain the approximation[6,10]

$$\Delta E_a^* \simeq (1 - m)\Delta E_a^*(C_2) + (n - 2 + 2m)[\Delta E_a^*(C_2) - \Delta E_a^*(C_1)] \quad (6.10)$$
$$+ \lambda_1 \sum N_{CC}\delta_C + \lambda_2[(n - 2 + 2m)\delta_C(C_1) + \sum \delta_C]$$

which is most useful in calculations of atomization energies because the lengthy separate calculation of nonbonded energies is not then required. In this equation, carbon atomic charges are already conveniently expressed in terms of ^{13}C chemical shifts and $\delta_C(C_1)$, for methane, is taken at -8 ppm from ethane.

Now we know that nonbonded energies differ somewhat from case to case in comparisons between structural isomers. On the other hand, the $\sum N_{CC}\delta_C$ and $\sum \delta_C$ terms are also structure-dependent. For this reason, a slight empirical readjustment of the parameters λ_1 and λ_2 compensates in part for the error introduced in assuming exact additivity for the nonbonded contributions. The recommended values are[6], in kcal mol^{-1} ppm^{-1} units, $\lambda_1 = 0.03244$ and $\lambda_2 = 0.05728$. With $\Delta E_a^*(C_2) = 710.54$ and $\Delta E_a^*(C_1) = 419.27$ kcal mol^{-1}, Eq. 6.10 becomes

TABLE 6.3. Approximate Energy Calculations of Saturated Hydrocarbons Using Eq. 6.11 without Explicit Consideration of Nonbonded Interactions

Molecule	$\Sigma N_{CC}\delta$	$\Sigma\delta$	ΔE_a^*, kcal mol^{-1}	
			Calcd.	Exptl.
Methane	0	−8	419.27	419.24
Ethane	0	0	710.54	710.54
Propane	39.8	29.6	1004.34	1004.07
Butane	91.0	52.8	1298.14	1298.15
Isobutane	113.4	74.8	1300.13	1299.70
Pentane	139.6	77.6	1591.95	1592.20
Isopentane	161.6	87.7	1593.24	1593.43
Neopentane	190.4	124.4	1596.28	1595.94
Hexane	188.4	102.2	1885.75	1885.95
2-Methylpentane	208.9	114.2	1887.11	1886.86
3-Methypentane	209.5	101.1	1886.37	1886.27
2,2-Dimethylbutane	231.4	127.1	1888.58	1888.86
2,3-Dimethylbutane	222.8	110.0	1887.32	1887.14
Heptane	233.0	124.3	2179.28	2179.81
2,2,3-Trimethylbutane	290.1	145.7	2182.36	2181.90
Cyclohexane	261.6	130.8	1760.85	1760.82
Methylcyclohexane	349.8	169.7	2056.75	2057.13
trans-Decalin	630.8	277.1	2815.54	2815.50
Adamantane	662.3	285.4	2688.12	2688.05
Bicyclo[2.2.2]octane	363.7	163.1	2218.72	2218.40

$$\Delta E_a^* \simeq 710.54(1 - m) + 290.812(n - 2 + 2m) + 0.03244\sum N_{CC}\delta_C \tag{6.11}$$
$$+ 0.05728\sum\delta_C \text{ kcal mol}^{-1}.$$

Equation 6.11 enables a rapid calculation of atomization energies, with an average error of ~ 0.22 kcal mol^{-1}, as indicated by the results given in Table 6.3. This error is still within experimental uncertainties and the loss in precision is a small price to pay for the considerable simplification in actual calculations. With Eq. 6.11 we have now a handy formula for practical use, whereas the detailed calculations involving explicitly nonbonded interactions are rather intended to demonstrate (with the minimum of compromise) the quality of our theoretical approach, Eqs. 5.46–5.48, at the best possible level, that of experimental accuracy. Numerous applications of Eq. 6.11 are presented in the next Chapter; but let us first spend a little more time with the results we have obtained so far.

8. CONCLUSIONS

The bottom line of the calculations presented in this Chapter for saturated hydrocarbon illustrates at least one merit of the theory described in Chapter

5: *the theoretical results are, indeed, at the level of experimental accuracy*. The comparisons between theory and experiment, while limited in number (mainly because of the scarcity of experimental vibrational energy results), are certainly significant because they were made on the basis of atomization energies ΔE_a^* deduced exclusively from experimental data, namely, the appropriate ZPE, H_T-H_0, and ΔH_f° energies.

Nonbonded Coulomb interaction energies, which turn out to be small when summed over the entire individual molecules, were estimated for use in the equation

$$\Delta E_a^* = \sum \varepsilon_{ij}^\circ + \sum_i \sum_j a_{ij} \Delta q_i - E_{nb}^*,$$

thus permitting comprehensive comparisons between theory and experiment at the best possible level. The average deviation (0.16 kcal mol^{-1}) is certainly well within experimental uncertainties. It has appeared that *differences in nonbonded Coulomb interactions cannot be made responsible for energy differences between structural isomers. It is the $\sum_i \sum_j a_{ij} \Delta q_i$ part, featuring the effects due to changes in net atomic charges, which is by far the leading term accounting for the energetic differences between isomers or conformers.* An approximation (Eq. 6.10, *viz.* 6.11) not requiring the tedious explicit calculation of nonbonded interactions suffers only a small loss in precision (average deviation $\simeq 0.22$ kcal mol^{-1}), still within experimental uncertainties. This approximation, which represents a considerable simplification, is exploited in Chapter 7, leading eventually to convenient simple rules for evaluating reliable ZPE + H_T-H_0 energies for saturated hydrocarbons, except in situations of extreme steric crowding.

An important result is that the *definition of atomic charges satisfying the energy expressions (Eqs. 5.46–5.48) is the same as that satisfying the* ^{13}C *NMR shift–charge correlation for sp^3 hybridized carbon atoms.* This means that valuable information concerning molecular energies can now be deduced simply from ^{13}C NMR spectra. Of course, caution must be exerted, as explained further below. For the saturated hydrocarbons, it is easy to derive the appropriate Δq_C's from ^{13}C shifts ($\Delta q_C = -0.148\delta_C$, Eq. 4.10) for use in the equation $\varepsilon_{CC} = \varepsilon_{CC}^\circ + a_{CC}\Delta q_{Ci} + a_{CC}\Delta q_{Cj}$. From charge normalization ($\Sigma q_H = -\Sigma q_C$) it is equally easy to deduce an appropriate average Δq_H for use in the equation $\varepsilon_{CH} = \varepsilon_{CH}^\circ + a_{CH}\Delta q_C + a_{HC}\Delta q_H$. Bond-by-bond calculations of this sort can always be kept at this elementary level, as illustrated in the detailed examples worked out in Section 2 for boat and chair cyclohexane. Finally, the energy of atomization is obtained from the sum of all the ε_{ij}'s, plus a small nonbonded contribution. While simple, this bond-by-bond approach can be somewhat lenghty at times.

In turn, the use of general formulas for $\sum_i \sum_j a_{ij} \Delta q_i$ and ΔE_a^* [like Eqs. 6.5, 6.8 and, more importantly, their approximation (Eq. 6.11), not requiring the separate calculation of nonbonded interactions] expedites considerably the evaluation of the ΔE_a^*'s from ^{13}C NMR shifts in homologous series of compounds. Moreover, and this is an important aspect, these formulas facilitate general conclusions to be drawn, valid for all the members of the class of compounds which is concerned. (Note that these remarks apply not only to the saturated hydrocarbons investigated here, but also to other series of compounds, e.g., the ethylenes described in Chapter 8). The $\lambda_1 \Sigma N_{CC} \delta_C + \lambda_2 \Sigma \delta_C$ term plays the leading role in the calculation of saturated hydrocarbons and differentiates isomers from one another. Incidentally, this term will be shown to appear in the calculation of any molecule containing alkyl substituents. Its precise evaluation is, therefore, of utmost importance in all cases involving large alkyl parts. In kcal mol^{-1} ppm^{-1} units, both λ_1 and λ_2 are positive quantities. Since ^{13}C shifts of saturated hydrocarbon-C atoms are farther downfield as carbon atoms are closer to electroneutrality (i.e., have gained electronic charge), we can express the overall conclusion as a rule of thumb: "*in comparisons between isomers, the more stable compound is that whose carbon skeleton best approaches electroneutrality, i.e., gains electronic charge from the hydrogen atoms, which is reflected by larger (downfield) δ_C values*". This, of course, is the echo of a conclusion reached earlier (Chapter 5.9) describing alkyl hydrogen atoms as electron reservoirs releasing electrons toward carbon in isomerizations leading to more stable compounds.

The present numerical solutions are directly applicable to normal and branched paraffins, cyclohexane, and polycyclic hydrocarbons constructed from chair and/or boat cyclohexane rings. Not unexpectedly, they fail to apply to other ring systems. Indeed, *ab initio* charge analyses (Table 4.8) indicate that $2p$ populations at CH_2 carbon atoms vary significantly with ring size, quite unlike the CH_2 carbons in the alicyclic and cyclohexanic series which exhibit constant populations at the $2p$ level. The appropriate a_{ij}'s and the charge–^{13}C shift correlation would, therefore, not any longer be those used in the numerical calculations presented so far. This means that there is still much room left for imaginative further studies, namely, in strained systems. At this stage, however, we may as well turn the page and launch into something useful with what we already know.

REFERENCES

1. T.L. Cottrell, "The Strength of Chemical Bonds", Academic Press, New York, NY, 1958.
2. D.K. Dalling and D.M. Grant, *J. Am. Chem. Soc.*, **96**, 1827 (1974).
3. J.D. Cox and G. Pilcher, "Thermochemistry of Organic and Organometallic Compounds", Academic Press, New York, NY, 1970.
4. J.P. Huvenne, G. Vergoten, G. Fleury, S. Odiot, and S. Fliszár, *Can. J. Chem.* **60**, 1347 (1982).
5. S. Fliszár, *J. Am. Chem. Soc.*, **102**, 6946 (1980).
6. H. Henry, G. Kean, and S. Fliszár, *J. Am. Chem. Soc.*, **99**, 5889 (1977).

7. T.L. Allen, *J. Chem. Phys.*, **31**, 1039 (1959).
8. F.A. Anet, C.H. Bradley, and G.W. Buchanan, *J. Am. Chem. Soc.*, **93**, 258 (1971).
9. A. Warshel and S. Lifson, *J. Chem. Phys.*, **53**, 582 (1970); J. Reisse, "Conformational Analysis, Scope and Present Limitations", G. Chiurdoglu, Ed., Academic Press, New York, NY, 1971.
10. S. Fliszár and M.-T. Béraldin, *Can. J. Chem.*, **57**, 1772 (1979).

CHAPTER 7

On the Role of Vibrational Energies

1. INTRODUCTION

The molecular energy of isolated molecules in their hypothetical vibrationless state is the primary quantity for the study of energy partitioning. Indeed, at this level the problem is not obscured by other forms of molecular energies, namely zero-point (ZPE) and thermal $(H_T\text{-}H_0)$ energies, which cannot be regarded as truly additive properties. Full advantage was taken of these facts, and appropriate equations were derived which give access to atomization energies of hypothetical vibrationless molecules at 0 K. On the other hand, standard enthalpies of formation, ΔH_f° (298.15, gas), provide valuable information about molecules containing vibrational energy. Hence, a combination of these two types of information allows us to evaluate ZPE + $H_T\text{-}H_0$ energies.

Such an approach is, of course, meaningful only when the energy of vibrationless molecules can be estimated with sufficient accuracy. So far, we know this to be the case with saturated hydrocarbons $C_nH_{2n+2-2m}$ containing m ($\geqslant 0$) six-membered rings. This class of compounds is, therefore, investigated first in order to collect sufficient information about vibrational energies, which would permit the construction of a simple empirical scheme for their evaluation. Extension to other classes of compounds is contemplated subsequently. Finally, using calculated ΔE_a^* atomization energies and empirically constructed ZPE + $H_T\text{-}H_0$ energies, it becomes possible, in a number of cases, to evaluate reasonably accurate enthalpies of formation.

2. OUTLINE OF CALCULATIONS

For the $C_nH_{2n+2-2m}$ hydrocarbons under study, accurate ΔE_a^* energies can be calculated from carbon–13 nuclear magnetic resonance shifts δ_C (ppm from ethane) by means of Eq. 6.11 (which was shown to agree with experimental ΔE_a^* values, Eq. 5.5, within experimental uncertainty). We can now equate these two expressions for ΔE_a^*. Inserting the experimental values (kcal mol^{-1} at $T = 298.15$ K) $\Delta H_f^\circ(C) = 170.89$, $\Delta H_f^\circ(H) = 52.09$, $\Delta E_a^*(C_2H_6) = 710.54$, and $\Delta E_a^*(CH_4) = 419.27$, as well as $\lambda_1 = 0.03244$, $\lambda_2 = 0.05728$ kcal mol^{-1} ppm^{-1}, and $\delta_C(CH_4) = -8$ ppm, the energy formula obtained[1] is (in kcal mol^{-1} at 298.15 K)

$$ZPE + H_T\text{-}H_0 = 27.692(1 - m) + 20.184n + 0.03244\sum N_{CC}\delta_C$$
$$+ 0.05728\sum\delta_C + \Delta H_f^\circ \tag{7.1}$$

which enables a rapid calculation of $ZPE + H_T\text{-}H_0$ energies from standard enthalpies of formation and C–13 NMR spectra. Of course, Eq. 7.1 is as accurate as Eq. 6.11 which has been tested earlier[1]. The possibility of deriving $ZPE + H_T\text{-}H_0$ energies in a simple manner by means of Eq. 7.1 offers now a new way of recognizing possible structure-related regularities from an examination of numerous cases. The results are as follows.

3. NONCYCLIC ALKANES

To begin with, we apply Eq. 7.1 to all the noncyclic alkanes (through C_9) which are either linear or contain only CH_3 branching but no quaternary C atoms. The idea behind this selection is to avoid any case involving an *a priori* visible steric hindrance of internal rotations. The results obtained in this way (Table 7.1) are well reproduced by the simple empirical additivity scheme[1]

$$ZPE + H_T\text{-}H_0 = 11.479 + 18.213n_1 + 17.870n_2 \tag{7.2}$$

where n_1 is the number of CH_2 groups forming the linear chain and n_2 is the number of CH_3 branchings.

Indeed, Table 7.1 indicates, firstly, that the $ZPE + H_T\text{-}H_0$ energies derived from the theoretical equation 7.1 are in excellent agreement with their experimental counterparts and, secondly, that these results are well reproduced by the empirical equation 7.2, which gives the $ZPE + H_T\text{-}H_0$ values with an average deviation of 0.125 kcal mol^{-1}. The change in $ZPE + H_T\text{-}H_0$ associated with methyl substitutions giving quaternary carbon atoms, however, can only be estimated as a rough average[1], $\sim 17.55 \pm 0.25$ kcal mol^{-1}, and should therefore be used with due care.

The practical aspects of Eq. 7.2 are clear: $ZPE + H_T\text{-}H_0$ energies can now be readily estimated and, together with the ΔE_a^*'s derived from ^{13}C NMR spectra (Eq. 6.10), they give access to accurate enthalpies of formation

TABLE 7.1. ZPE + H_T-H_0 Energies of Selected Paraffins Containing No Quaternary Carbon Atoms

| Molecule | ZPE + H_T-H_0, kcal/mol | | |
	Exptl.	Eq. 7.1	Eq. 7.2
C_1	29.5	29.53	29.69
C_2	48.02	48.02	47.91
C_3	65.94	66.20	66.12
C_4	84.38	84.37	84.33
2-MeC_3	83.89	84.32	83.99
C_5	102.83	102.59	102.54
2-MeC_4	102.14	101.96	102.20
C_6	120.99	120.78	120.76
2-MeC_5		120.66	120.41
3-MeC_5		120.56	120.41
2,3-Me$_2$$C_4$	119.65	119.90	120.07
C_7	139.29	138.93	138.97
2-MeC_6		138.65	138.63
2,3-Me$_2$$C_5$		138.05	138.28
2,4-Me$_2$$C_5$		138.25	138.28
C_8		157.16	157.18
2,3-Me$_2$$C_6$		156.87	156.50
2,4-Me$_2$$C_6$		156.52	156.50
2,3,4-Me$_3$$C_5$		156.23	156.15
C_9		175.35	175.40
4-MeC_8		175.03	175.05
2,3,5-Me$_3$$C_6$		174.26	174.37

Zero-point energies were calculated from Eq. 5.6 using the frequencies given in: R.G. Snyder and J.H. Schachtschneider, Ref. 2. The H_T-H_0 values are from Ref. 3. The experimental ^{13}C and ΔH_f° data used in Eq. 7.1 are indicated in Table 7.2.

(Eq. 5.5). The results are indicated in Table 7.2, for the compounds containing no quaternary carbon atoms. The agreement between calculated and experimental ΔH_f° values (0.14 kcal mol^{-1} average error) certainly supports the theory, namely Eqs. 5.46–5.48, which led to these results, and the validity of the approximation expressed by Eq. 7.2. The fact that the λ_1 and λ_2 parameters used here differ slightly from the strictly theoretical ones (for the reasons explained in Chapter 6.7) is of no consequence in this respect. Indeed, the same agreement is obtained for a selection of 14 of these compounds with the use of theoretical λ_1 and λ_2 values provided, of course, that nonbonded interactions are explicitly taken into account (see Table 6.1).

With no exception, all calculations considered so far have been carried out following an explicit separation of vibrational energies from the nonvibrational part. It is now instructive to examine the possibility of con-

TABLE 7.2. Comparison of ΔH_f°'s Calculated from Eqs. 7.1 and 7.2 or Eq. 7.4 with Experimental Results

			$-\Delta H_f^\circ$, kcal/mol		
Molecule	$\Sigma N_{CC}\delta$	$\Sigma\delta$	Exptl.	Eqs. 7.1, 7.2	Eq. 7.4
C_1	0	−8.0	17.89	17.73	18.15
C_2	0	0	20.04	20.15	20.08
C_3	39.8	29.6	25.02	25.11	24.97
C_4	91.0	52.8	30.03	30.07	30.03
2-MeC$_3$	113.4	74.8	32.07	32.40	32.22
C_5	139.6	77.6	35.00	35.05	35.05
2-MeC$_4$	161.6	87.7	36.92	36.68	36.59
2,2-Me$_2$C$_3$	190.4	124.4	40.27		39.85
C_6	188.0	102.0	39.96	39.98	40.05
2-MeC$_5$	211.8	116.1	41.66	41.91	41.88
3-MeC$_5$	212.5	102.9	41.02	41.17	41.21
2,3-Me$_2$C$_4$	223.8	110.6	42.49	42.32	42.13
C_7	235.4	125.7	44.89	44.85	44.97
2-MeC$_6$	257.5	138.2	46.60	46.62	46.64
3-MeC$_6$	255.5	126.7	45.96	45.90	45.93
2,3-Me$_2$C$_5$	267.2	123.2	46.65	46.42	46.27
2,4-Me$_2$C$_5$	274.6	151.3	48.30	48.27	48.11
2,2-Me$_2$C$_5$	288.4	159.7	49.29		49.18
2,2,3-Me$_3$C$_4$	290.1	145.7	48.96		48.51
C_8	284.2	150.0	49.82	49.80	49.98
2-MeC$_7$	306.5	162.8	51.50	51.59	51.67
3-MeC$_7$	307.4	153.0	50.82	51.06	51.19
4-MeC$_7$	303.4	153.6	50.69	50.96	51.04
2,3-Me$_2$C$_6$	315.8	150.0	51.13	51.50	51.41
2,4-Me$_2$C$_6$	321.1	163.8	52.44	52.46	52.39
2,5-Me$_2$C$_6$	327.8	174.6	53.21	53.30	53.27
3,4-Me$_2$C$_6$	314.2	139.0	50.91	50.82	50.75
2,3,4-Me$_3$C$_5$	324.3	148.7	51.97	52.05	51.72
2,2-Me$_2$C$_6$	331.7	180.8	53.71		53.77
3,3-Me$_2$C$_6$	319.1	159.0	52.61		52.04
2,2,3-Me$_3$C$_5$	344.8	164.8	52.61		53.51
2,3,3-Me$_3$C$_5$	319.0	145.7	51.73		51.32
2,2,3,3-Me$_4$C$_4$	353.8	178.0	53.99		54.63
C_9	332.0	174.1	54.54	54.69	54.93
4-MeC$_8$	350.7	177.3	56.19	55.83	55.95
2,3,5-Me$_3$C$_6$	378.0	185.4	57.97	57.86	57.62

The ^{13}C data were taken from Ref. 4 (C_1, C_2), Ref. 5 (C_3—C_6), and Ref. 6 (C_7—C_9). Except as noted below, the experimental ΔH_f° values are from Ref. 3. The values for C_2, C_3, and 2-MeC$_3$ are taken from Ref. 7, and that of 2,2-Me$_2$C$_3$ from Ref. 8. The values for 2,3-Me$_2$C$_6$ and for the C_9 alkanes are taken from Ref. 9. The present selection of experimental ΔH_f° values is not intended to represent a critical collection of "best values". Slightly different results are reported in Ref. 9a for C_2 (−20.24), C_3 (−24.82) and C_4 (−30.15 kcal/mol), all other values being those given in this table. Similarly, one finds (Ref. 9b), in kcal/mol, −35.07 for C_5, −36.55 (2-MeC$_4$) and −40.09 (2,2-Me$_2$C$_3$). Experimental uncertainties of this magnitude do not affect the general conclusions drawn in this work. The parameters of Eq. 7.4 are $\lambda_1 = 0.04531$ and $\lambda_2 = 0.05338$.

structing a scheme leading directly to enthalpy values. Of course, the construction of an additivity scheme involving vibrational energies is, in itself, not free from conceptual difficulties and is considered here only on an empirical basis. It is also true, however, that the problems raised in this manner are no more acute than those hidden in popular additivity schemes involving directly ΔH_a or ΔE_a, because any such scheme implicitly distributes vibrational energies by treating them *de facto* as if they were additive. This point can be shown as follows, by considering the consequences of forcing exact additivity of vibrational energies. Let us temporarily assume that one single value (e.g., 18.213 kcal mol^{-1}) describes the increment in $ZPE + H_T-H_0$ for one added CH_2 group, regardless of the point of methyl substitution, i.e.,

$$ZPE + H_T-H_0 = 11.479 + 18.213n = 47.905 + 18.213(n-2)$$

where 47.905 kcal mol^{-1} is the value for ethane. (This equation is a form of Eq. 6.9 defining exact additivity). Then, using Eqs. 5.1, 5.5, and 6.10, we obtain the following expressions[1]

$$\Delta H_a = \Delta H_a(2) + (n-2)[\Delta H_a(2) - \Delta H_a(1)] + \lambda_1\sum N_{cc}\delta_c$$
$$+ \lambda_2[(n-2)\delta_c(1) + \sum\delta_c] \tag{7.3}$$

$$\Delta H_f^\circ = \Delta H_f^\circ(2) + (n-2)[\Delta H_f^\circ(2) - \Delta H_f^\circ(1)] - \lambda_1\sum N_{cc}\delta_c$$
$$- \lambda_2[(n-2)\delta_c(1) + \sum\delta_c] \tag{7.4}$$

in which the indices 1 and 2 refer to methane and ethane, respectively. These equations describe in fact a bond-enthalpy additivity scheme, $\Delta H_a = \sum H(\text{bond})$ with individual bond contributions depending on the charges of the bond-forming atoms, implying exact additivity not only of nonbonded interactions, but also of the vibrational energy contributions. Now we know that both nonbonded and vibrational energies do depend on structural features, which is also the case with the $\sum N_{cc}\delta_c$ and $\sum\delta_c$ terms. Equations 7.3 and 7.4 give generally satisfactory results (Table 7.2) because an empirical readjustment of the parameters λ_1 and λ_2 compensates for the errors introduced in assuming exact additivity of the nonbonded and the vibrational energy terms. Hence, the success of Eqs. 7.3 and 7.4 is in part due to fortunate compensations and should not be interpreted as a proof for postulated intrinsic qualities of bond-enthalpy additivity schemes. The appropriate parameters for use in Eqs. 7.3 and 7.4 are $\lambda_1 = 0.04531$ and $\lambda_2 = 0.05338$ kcal mol^{-1} ppm^{-1}, and the average deviation between calculated and experimental values is now 0.25 kcal mol^{-1}.

A more instructive way to look at the results is the following. If we agree upon calling "steric effect" any departure from exact additivity, as defined by Eq. 6.9, it appears from Eq. 7.3 that

$$\text{"Steric effect"} = \lambda_1\sum N_{cc}\delta_c + \lambda_2[(n-2)\delta_c(1) + \sum\delta_c].$$

Of course, when referring to enthalpies (e.g., at 25°C), this expression is only an approximate one because Eq. 7.3 itself is an approximation. At the vibrationless level, however, this definition of "steric effects" is strictly valid for ΔE_a^{*bonds} (see Eq. 6.6) and requires then the use of theoretical λ_1 and λ_2 values, i.e., 0.0356 and 0.0529 kcal mol^{-1} ppm^{-1}, respectively. For ΔE_a^*, which includes nonbonded interactions, and even more so for enthalpies, λ_1 and λ_2 are asked to correct in an empirical manner for any departure from exact additivity as regards nonbonded, $viz.$, nonbonded and vibrational contributions. Still then, the numerical values of the "adjusted" λ_1 and λ_2 parameters indicate that by far the largest part of the steric effects originates in the structure dependence of charge distributions and the way local electron populations govern bond energies.

4. CYCLOALKANES

We can now turn to the study of cyclic hydrocarbons and examine to what extent the additivity scheme describing ZPE + H_T-H_0 energies in acyclic compounds (Eq. 7.2) is applicable to cyclic structures. To begin with, we select molecules which are free from effects such as butane $gauche$ or $skew$ pentane interactions.

Firstly, we note that a CH_2 group contributes 18.213 kcal mol^{-1} if no tertiary C atom is formed and 17.870 kcal mol^{-1} when branching occurs to give a tertiary C atom. We construct tentatively the ZPE + H_T-H_0 energies of cycloalkanes by counting 18.213 kcal mol^{-1} for each C atom of the molecule (rule 1) and subtract $18.213 - 17.870 = 0.343$ kcal mol^{-1} for each tertiary C atom that is formed (rule 2). A number of H atoms are removed in the formation of polycyclic compounds, e.g., 2 H atoms in $trans$-decalin. The 11.479 kcal mol^{-1} term of Eq. 7.2 represents roughly the contribution of two H atoms. Accordingly, we approximate the decrease in ZPE + H_T-H_0 due to the loss of one H atom by subtracting $11.479/2 = 5.740$ kcal mol^{-1} per H atom that is removed (rule 3). These rules express clearly the fact that up to this point cyclic compounds are considered precisely on the same basis as the acyclic ones, following strictly the description given by Eq. 7.2. Considering now that Eq. 7.2 describes noncyclic compounds in which internal rotations are as free as possible and, moreover, that internal rotations are hindered in cyclic compounds, we subtract tentatively $RT/2 = 0.296$ kcal mol^{-1} (at 298.15 K) for each CC bond in the cycle (rule 4). The results are presented in Table 7.3, together with results derived from Eq. 7.1.

The comparison of the ZPE + H_T-H_0 energies deduced from Eq. 7.1 with their experimental counterparts clearly confirms the validity of Eq. 7.1 for this class of compounds and adds credibility to results derived in this fashion. On the other hand, the remarkable agreement between ZPE + H_T-H_0 energies constructed from the admittedly crude rules 1–4 and their counterparts deduced from Eq. 7.1 suggests that these cycloalkanes behave in essence like acyclic alkanes inasmuch as vibrational energies are concerned, provided that the suppression of internal rotations is adequately taken into account.

TABLE 7.3. ZPE + H_T-H_0 Energies of Selected Cycloalkanes

| Molecule | ZPE + H_T-H_0, kcal mol^{-1} | | |
	Exptl.	Eq. 7.1	Rules 1–4
Cyclohexane	107.54	107.58	107.50
Methylcyclohexane	125.73	125.37	125.37
Ethylcyclohexane		144.17	143.59
n-Butylcyclohexane		180.30	180.01
cis-1,3-Dimethylcyclohexane		143.46	143.24
trans-1,4-Dimethylcyclohexane		143.46	143.24
1-cis-3-cis-5-Trimethylcyclohexane		161.54	161.11
trans-Decalin	166.90	166.96	166.71
trans-syn-trans-Perhydroanthracene		225.64	225.90
trans-anti-trans-Perhydroanthracene		225.96	225.90
Adamantane	153.54	153.64	154.25
Twistane		153.68	154.25
Bicyclo[2.2.2]octane	130.83	131.17	130.87

The value for bicyclo[2.2.2]octane is given in Ref. 10. The other results are given in Ref. 11 and were deduced from the spectroscopic data of Ref. 2 and the H_T-H_0 values reported in Ref. 3. The experimental ^{13}C and ΔH_f° data used in Eq. 7.1 are indicated in Table 7.4.

Failure to do so would result in errors which could be mistaken for "ring strain", e.g., 1.8 kcal mol^{-1} for cyclohexane and 3.6 kcal mol^{-1} for adamantane, although these are precisely the type of systems for which ring strain should not exist. The fact that the $RT/2$ correction appears to be adequate further suggests that in acyclic hydrocarbons the contributions from internal rotations seem to be fairly close to $RT/2$, even in relatively long chains.

Now, of course, there is no reason to assume that the individual energy contributions in cycloalkanes are necessarily *exactly* the same as in noncyclic paraffins. To begin with, the comparison of ZPE + H_T-H_0 energies derived by means of Eq. 7.1 for isomer pairs differing by the number of butane *gauche* interactions (e.g., *cis-* vs. *trans*-decalin) indicates a lowering of ZPE + H_T-H_0 energy by 0.853 kcal mol^{-1} for one *gauche* interaction. A multiple regression analysis carried out using ZPE + H_T-H_0 energies derived for sixteen cycloalkanes containing no quaternary C atoms indicates, firstly, a contribution of 18.249 kcal mol^{-1} for each carbon atom and, secondly, a decrease of 0.322 kcal mol^{-1} for each tertiary C atom of the molecule. Moreover, the removal of one H atom in the formation of polycyclic structures reduces the ZPE + H_T-H_0 energy by 5.925 kcal mol^{-1}. Finally, this energy is further lowered by 0.296 (= $RT/2$) kcal mol^{-1} for each CC bond of the cycle(s). These results, which resemble closely the rules inferred from noncyclic alkanes, are expressed by Eq. 7.5[1]

$$ZPE + H_T\text{-}H_0 = 11.850(1 - m) + 18.249n - 0.322n_{\text{tert}}$$
$$- \frac{1}{2}(n_C + m - 1)RT - 0.853n_g \tag{7.5}$$

TABLE 7.4. Calculated ΔE_a^*, ZPE $+ H_T\text{-}H_0$ and ΔH_f° (298.15, gas) Energies (kcal/mol) of Cycloalkanes

Molecule	$\Sigma N_{CC}\delta$	$\Sigma\delta$	ΔE_a^*	ZPE $+ H_T\text{-}H_0$	$-\Delta H_f^\circ$ Calcd.	$-\Delta H_f^\circ$ Exptl.
1 Cyclohexane	261.6	130.8	1760.86	107.72	29.36	29.50
2 Methylcyclohexane	349.8	169.7	2056.77	125.65	36.71	36.99
3 Ethylcyclohexane	402.1	186.8	2350.26	143.89	41.33	41.05
4 n-Butylcyclohexane	492.1	234.3	2937.52	180.39	50.83	50.92
5 cis-1,3-Dimethylcyclohexane	437.0	208.5	2352.63	143.57	44.02	44.13
6 trans-1,3-Dimethylcyclohexane	373.9	180.4	2348.97	141.87	42.06	42.20
7 trans-1,4-Dimethylcyclohexane	436.8	208.5	2352.62	143.57	44.01	44.12
8 cis-1,4-Dimethylcyclohexane	378.7	179.3	2349.07	141.87	42.16	42.22
9 1,1-Dimethylcyclohexane	391.4	194.5	2350.35	142.19	43.12	43.26
10 1,1,3-Trimethylcyclohexane	483.6	237.7	2646.63	160.12	50.84	
11 1,1,4-Trimethylcyclohexane	474.8	231.1	2645.97	160.12	50.18	
12 1-cis-3-cis-5-Trimethylcyclohexane	527.3	248.9	2648.69	161.50	51.52	51.48
13 1-cis-3-trans-5-Trimethylcyclohexane	462.6	223.1	2645.11	159.79	49.65	49.37
14 trans-Decalin	630.8	277.1	2815.56	166.74	43.74	43.52
15 cis-Decalin	527.0	232.5	2809.64	164.18	40.38	40.43
16 trans-anti-1-Methyldecalin	717.9	308.2	3110.98	183.81	51.46	
17 cis-syn-1-Methyldecalin	606.0	259.4	3104.56	181.25	47.60	
18 trans-syn-2-Methyldecalin	712.6	313.7	3111.13	184.67	50.75	
19 cis-syn-2-Methyldecalin	620.2	273.8	3105.84	182.11	48.02	
20 trans-9-Methyldecalin	637.6	274.7	3106.46	181.58	49.17	
21 cis-9-Methyldecalin	590.7	261.6	3104.17	180.72	47.76	
22 trans-syn-2-syn-7-Dimethyldecalin	797.1	351.8	3406.87	202.59	57.94	
23 trans-anti-1-syn-8-Dimethyldecalin	709.1	301.3	3401.12	199.18	55.60	
24 trans-anti-1-syn-3-Dimethyldecalin	794.4	342.8	3406.26	201.74	58.18	

25	*trans-syn-trans*-Perhydroanthracene	997.4	3870.15	225.76	58.01	58.13
26	*trans-anti-trans*-Perhydroanthracene	900.5	3865.08	225.76	52.94	52.74
27	*cis-trans*-Perhydroanthracene	890.1	3864.17	223.20	54.59	
28	*cis-anti-cis*-Perhydroanthracene	779.6	3857.98	220.64	50.96	
29	*trans-anti-trans*-Perhydrophenanthrene	977.9	3868.66	224.91	57.37	
30	*trans-anti-cis*-Perhydrophenanthrene	872.0	3862.85	222.35	54.12	
31	*trans-syn-cis*-Perhydrophenanthrene	879.0	3863.22	222.35	54.49	
32	Bicyclo[2.2.2]octane	363.7	2218.74	130.83	24.09	23.75
33	Adamantane	662.3	2688.15	153.95	30.34	30.65
34	Twistane	514.4	2679.14	153.95	21.36	21.6
35	Spiro[5.5]undecane	552.2	3102.24	181.60	44.93	44.81

The carbon-13 NMR results are extracted from Ref. 12 (**1, 2, 5–13**), Ref. 13 (**3, 4**), Ref. 14 (**14–24, 35**), Ref. 15 (**25–31**), Ref. 16 (**32**), Ref. 17 (**33**), and Ref. 18 (**34**). The experimental enthalpies of formation are taken from Ref. 19 (**1–8, 14, 15, 25, 26**), Ref. 3 (**9**), Ref. 20 (**12, 13**, from equilibration experiments), Ref. 10 (**32, 33**), Ref. 21 (**34**) and Ref. 22 (**35**).

where n_C and $n_C + m - 1$ are, respectively, the numbers of C atoms and CC bonds in the cycle(s) and $2(m - 1)$ is the number of H atoms removed in the formation of polycyclic compounds. The number of butane *gauche* interactions is n_g. Of course, not too much importance should be attached to the precise values of these parameters as they are subject to slight changes depending upon whether or not one or another compound is left out in this type of analysis.

This approximation yields accurate estimates of ZPE + H_T-H_0 energies, which can be compared with those of Table 7.3, and involves only the knowledge of the molecular structures. Indeed, enthalpies of formation calculated from Eq. 7.1 (or Eqs. 5.5 and 6.10) using now the results given by Eq. 7.5 and the appropriate ^{13}C NMR data are accurate within 0.17 kcal mol^{-1} (average deviation) when compared with their experimental counterparts. The results are given in Table 7.4. This agreement may be regarded as a further confirmation of the validity of the theory underlying these calculations, moreover as for a number of these compounds (Table 6.1) detailed calculations involving spectroscopic ZPE + H_T-H_0 energies and explicit nonbonded energies lead to the same result with the use of entirely theoretical λ_1 and λ_2 parameters.

It is noteworthy that the change from chair to boat cyclohexane has no effect (or only a very minor one) on ZPE + H_T-H_0. This is clearly indicated by the results obtained from *trans-anti-trans*-perhydroanthracene (**26**), bicyclo[2.2.2] octane (**32**), and twistane (**34**), and is further supported by the ZPE + H_T-H_0 energy determined for **32** by Boyd *et al.*[10], 130.83 kcal mol^{-1}, which is precisely the value derived from Eq. 7.5. Finally, force-field calculations of the fundamental frequencies of boat and chair cyclohexane confirm that the two forms differ by no more than 0.1 kcal mol^{-1} from one another in total vibrational energy[23].

Isomer pairs differing by the number of *gauche* interactions (e.g., *trans-vs. cis*-decalin, **5** *vs.* **6**, **7** *vs.* **8**, etc.) usually reveal a decrease of ~ 1.8 kcal mol^{-1} in ΔE_a^* energy for one *gauche* interaction, i.e., a destabilization of the chemical bonds due to a loss of electronic charge at the carbon atoms in favor of the hydrogens. Because of the concurrent decrease in vibrational energy by ~ 0.85 kcal mol^{-1}, which is a stabilizing effect, the total destabilization due to one *gauche* interaction is ~ 0.95 kcal mol^{-1}. Consider, for example, methyl-substituted cyclohexane isomers differing by the conformational position of one methyl group (i.e., **5** *vs.* **6**, **7** *vs.* **8**, **12** *vs.* **13**). The average stabilization of the equatorial *vs.* the axial form of ~ 1.9 kcal mol^{-1} (enthalpy, at 25°C) calculated from our scheme is reasonably close to the currently accepted value of ~ 1.8 kcal mol^{-1}, which itself is not an observed quantity but an estimate extrapolated from, and consistent with, numerous data collected from other molecules. This analysis of the *gauche* effect, separating the vibrationless from the vibrational part, clearly shows that it has nothing to do with Coulomb-type interactions between nonbonded atoms.

TABLE 7.5. Zero-point and H_T-H_0
Energies of Selected Olefins (kcal mol^{-1})

Molecule	ZPE	H_T-H_0
Ethene	30.83[a]	2.53
	30.54[b]	
Propene	48.58[c]	3.24
1-Butene	65.62[d]	4.11
cis-2-Butene	65.97[c]	3.94
trans-2-Butene	65.51[c]	4.19
Isobutene	65.57[c,e]	4.08
	65.41[f]	

The H_T-H_0 values are taken from Ref. 3.
The ZPE values are deduced from vibrational analyses given in a) Ref. 26, b) Ref. 27, c) Ref. 28, d) Ref. 29, e) Ref. 30, and f) Ref. 31.

In comparisons between isomers, the order of stability indicated by the ΔE_a^* energies is in most cases also reflected in the ΔH_f°'s. An interesting exception is the cis-trans-perhydroanthracene which is 0.91 kcal mol^{-1} less stable in ΔE_a^* energy than its trans-anti-trans isomer but more stable in terms of ΔH_f° by 1.65 kcal mol^{-1} because of its low content in vibrational energy. This point is confirmed by Allinger's equilibration experiments[24].

While this last example further stresses the important role of the vibrational contributions in determining molecular stabilities, it is clear that the present analysis is incomplete as it would not account for a number of possible structural features. In Table 7.4 we have only considered molecules for which the structural dependence of the ZPE + H_T-H_0 energies has been substantiated by an adequate body of experimental data, thus leaving out structural features (e.g., skew pentane, or interactions between methyl substituents in a 1,2 position) for which no, or too little, experimental information is presently available. Because of this selection, our predicted ΔH_f° values should represent reasonably good estimates. They are, indeed, supported by strong indirect evidence[1], from equilibration experiments[24,25].

5. ETHYLENIC HYDROCARBONS

Spectroscopic information is scarce for this class of compounds. The appropriate ZPE's, calculated from literature data using Eq. 5.6, are indicated in Table 7.5.

Invoking the results described for linear and branched alkanes (Eq. 7.2), namely the evidence that one CH_2 added to an alkyl group increases ZPE + H_T-H_0 by 18.213 kcal mol^{-1} whereas branching lowers this energy by 0.343

kcal mol^{-1}, it seems reasonable to assume a similar pattern for the normal and branched ethylenic hydrocarbons. This approximation is expressed by Eq. 7.6.

$$ZPE + H_T\text{-}H_0 \simeq 33.35 + 18.213(n - 2) - 0.343\, n_{\text{branching}}. \quad (7.6)$$

The experimental results given in Table 7.5 certainly support Eq. 7.6 although, evidently, additional vibrational energy data would be desirable. It is noteworthy, however, that numerous enthalpies of formation calculated from the appropriate ΔE_a^* atomization energies and Eqs. 5.5 and 7.6 are in satisfactory agreement with the corresponding experimental results suggesting that, as a whole, Eq. 7.6 represents a reasonable approximation.

6. CARBONYL COMPOUNDS

Complete vibrational analyses of carbonyl compounds are particularly scarce as far as the number of molecules is concerned. Results deduced from Eqs. 5.6 and 5.7 using experimental and calculated fundamental frequencies of acetaldehyde, acetone, and diethylketone are indicated in Table 7.6.

Using theoretical ΔE_a^* energies, it is, however, possible to estimate the increment in $ZPE + H_T\text{-}H_0$ energy for one added CH_2 group. This is simply done by means of Eq. 5.5 which allows comparisons involving experimental enthalpies of formation. In this way, an increment of ~ 18.3 kcal mol^{-1} can be assigned to each added CH_2 group, with respect to the closest "parent" compound whose spectroscopic $ZPE + H_T\text{-}H_0$ result is known. For example, the value for propanal ($36.65 + 18.3$) is estimated with respect to that of ethanal (36.65 kcal mol^{-1}) and the value for butanone (72.89) is estimated with respect to propanone (54.59 kcal mol^{-1}). The "final" $ZPE + H_T\text{-}H_0$ values derived in this fashion are indicated in Table 9.6, together with the corresponding detailed calculations of atomization energies.

This sort of "additive" behavior of $ZPE + H_T\text{-}H_0$ energies now appears to be plausible, considering the results of similar nature obtained for the hydrocarbons. Nevertheless, the necessity of estimating $ZPE + H_T\text{-}H_0$ energies in empirical fashion would appear to diminish the convincing power of the agreement between calculated and experimental energies in support of theory. It must be considered, however, that carbonyl compounds represent a class of molecules where the charge effects vary considerably from case to case, mainly because of the large variations in charge at the oxygen atom. Hence, the quality of the agreement depends primarily on the performance of the theoretical a_{ij} coefficients, namely on that of oxygen (Eq. 5.47). For this reason, the relatively minor uncertainties in the estimates of $ZPE + H_T\text{-}H_0$ energies should not overshadow the validity of the numerical tests which are offered in Chapter 9.4.

TABLE 7.6. Zero-point and Heat-Content Energies (kcal/mol) of Selected Carbonyl Compounds and Ethers

Molecule	ZPE	H_T-H_0	Estimated ZPE + H_T-H_0
CH_3CHO	33.58	3.07	
$(CH_3)_2CO$	50.59	4.00	
$(C_2H_5)_2CO$	86.68	4.84	
$(CH_3)_2O$	49.91	2.74	52.55 (−0.10)
$CH_3OC_2H_5$	65.75	4.11	70.19 (0.33)
$CH_3O-i-C_3H_7$	82.74	4.92	87.83 (0.17)
$(C_2H_5)_2O$	82.82	4.94	87.83 (0.07)
$C_2H_5O-i-C_3H_7$	99.59	5.81	105.47 (0.07)
$(i-C_3H_7)_2O$	116.65	6.67	123.11 (−0.21)
Tetrahydropyran	88.83	4.04	
1,4-Dioxane	74.61	3.87	

The results were deduced in the harmonic oscillator approximation (Eqs. 5.6 and 5.7) using the vibrational analyses of acetaldehyde and acetone (Ref. 32) and diethylketone (Ref. 33). For the ethers, the IR and Raman spectra are taken from Refs. 34 and 35, for dimethylether. The "estimated" values are deduced from the approximation ZPE + H_T-H_0 = 52.55 + 17.64($n-2$) kcal/mol, as indicated in the text; the differences between these estimates and those based on spectroscopic evidence are given in parentheses.

7. ETHERS

The ZPE and H_T-H_0 results deduced from the appropriate vibrational analyses are collected in Table 7.6. The results for dialkylethers are adequately represented by the approximation

$$ZPE + H_T-H_0 = 52.55 + 17.64(n-2) \text{ kcal mol}^{-1} \qquad (7.7)$$

where n is the number of carbon atoms. The average deviation between these estimates and those based on spectroscopic evidence is 0.16 kcal mol^{-1}. Detailed calculations involving these estimates, theoretical ΔE_a^* atomization energies, and experimental enthalpies of formation are satisfactory, with an average deviation between calculated and observed energies of 0.17 kcal mol^{-1} (Chapter 9.3).

8. ON THE QUASI-ADDITIVITY OF VIBRATIONAL ENERGIES

The vibrational energy $\Sigma F((v_i, T)$ corresponding to the fundamental frequencies v_i of a polyatomic molecule at some temperature T (Chapter 5.2) is, in its very essence, not a property which can be fairly apportioned among the individual chemical bonds of that molecule. It remains, however, that the accuracy of empirical additivity formulas representing $\Sigma F(v_i, T)$ exceeds what one could have hoped for *a priori*. This is the result of a number of fortunate internal compensations which are examined as follows, in an analysis revealing (as an interesting by-product) some merits of popular bond additivity schemes.

To begin with, let us describe linear saturated hydrocarbons C_nH_{2n+2} by means of Eq. 5.3 which introduces the appropriate nuclear-motion contributions. The energies of atomization, $\Delta E_{a,n}$ and $\Delta E_{a,n-1}$, of two consecutive normal alkanes C_nH_{2n+2} and $C_{n-1}H_{2n}$, respectively, are

$$\Delta E_{a,n} = \Delta E_{a,n}^* - \sum^{9n} F(v_i, T) + \frac{9n}{2} RT$$

$$\Delta E_{a,n-1} = \Delta E_{a,n-1}^* - \sum^{9n-9} F(v_i, T) + \frac{(9n-9)}{2} RT.$$

The difference

$$\Delta E_{a,n} - \Delta E_{a,n-1} = \Delta E_{a,n}^* - \Delta E_{a,n-1}^* - \sum^{9n} F(v_i, T) + \sum^{9n-9} F(v_i, T) + \frac{9}{2} RT.$$

(7.8)

between two consecutive members represents the energy contribution of one added CH_2 group, i.e., roughly speaking, of one CC plus two CH bonds. The equation itself is exact but the precise meaning of the "CC and CH bond energies", both at the vibrationless level and for the actual molecule at some temperature, T, remains to be clarified.

Temporarily leaving things as they are, let us now consider the difference

$$\Delta E_{a,n} - (n-1)(\Delta E_{a,n} - \Delta E_{a,n-1}) = 4(E_{CH})$$

where (E_{CH}) is a quantity "resembling a CH bond energy". The approximate validity of this metaphor can be recognized by observing that for $n = 2$, for example, $\Delta E_{a,n}$ represents the contributions of one CC and six CH bonds. Subtracting one CC and two CH bonds leaves us with four CH bonds. By means of Eq. 7.8 it is readily deduced that

$$(E_{CH}) = (E_{CH}^*) + \frac{1}{4}(n-2)\sum^{9n} F(v_i, T) - \frac{1}{4}(n-1)\sum^{9n-9} F(v_i, T) + \frac{9}{8} RT \quad (7.9)$$

where

$$(E_{CH}^*) = \frac{1}{4}[\Delta E_{a,n}^* - (n-1)(\Delta E_{a,n}^* - \Delta E_{a,n-1}^*)] \quad (7.10)$$

is the counterpart of (E_{CH}) for the molecule in its hypothetical vibrationless state. Similarly,

$$(n+1)(\Delta E_{a,n} - \Delta E_{a,n-1}) - \Delta E_{a,n} = 2(E_{CC})$$

yields a sort of "CC bond energy", i.e., from Eq. 7.8,

$$(E_{CC}) = (E^*_{CC}) - \frac{1}{2}n\sum_{}^{9n} F(v_i, T) + \frac{1}{2}(n + 1) \sum_{}^{9n-9} F(v_i, T) + \frac{9}{4}RT \quad (7.11)$$

where

$$(E^*_{CC}) = \frac{1}{2}[(n + 1)(\Delta E^*_{a,n} - \Delta E^*_{a,n-1}) - \Delta E^*_{a,n}] \quad (7.12)$$

refers to the vibrationless molecule at 0 K. All the equations used so far are exact. Now we must put their interpretation in a correct physical perspective, namely, as regards the quantities vaguely described as CC and CH bond energies.

This interpretation can be offered because of two facts disclosed by numerical analyses. It turns out, indeed, that both $\Delta E_{a,n} - \Delta E_{a,n-1} = 278.25 \pm 0.25$ and $\Delta E^*_{a,n} - \Delta E^*_{a,n-1} = 293.78 \pm 0.04$ kcal mol^{-1} are practically invariant.* As a consequence, $\sum^{9n} F(v_i, T) - \sum^{9n-9} F(v_i, T)$ is also constant (~ 18.2 kcal mol^{-1}, from Eq. 7.8)—an observation which is exploited further below. The "composition" of $\Delta E^*_{a,n} - \Delta E^*_{a,n-1}$ is readily deduced from Eq. 6.7, using the "apparent" CC and CH bond energies[36] (Chapter 6.5) which include major corrections contributing to charge normalization, i.e., $\varepsilon^{app}_{CC} = \varepsilon^\circ_{CC} - a_{HC}q^\circ_C/2 = 80.723$ and $\varepsilon^{app}_{CH} = \varepsilon^\circ_{CH} - a_{HC}q^\circ_H/4 = 104.958$ kcal mol^{-1}. The sum $\varepsilon^{app}_{CC} + 2\varepsilon^{app}_{CH}$ plus the average increase in $\lambda_1\sum N_{CC}\Delta q_C + \lambda_2\sum\Delta q_C$ for one added CH$_2$ (3.01 kcal mol^{-1})† and, finally, the corresponding increment in nonbonded energy (~ 0.13 kcal mol^{-1}) add up to give the value observed for $\Delta E^*_{a,n} - \Delta E^*_{a,n-1}$ and reveal the composite nature of this term. In spite of this, the *image* projected by the invariance of $\Delta E^*_a - \Delta E^*_{a,n-1}$ is that of an empirical bond additivity scheme involving seemingly constant CC and CH bond contributions, i.e., from Eq. 7.10 or 7.12, respectively,

$$\Delta E^*_{a,n} = 4(E^*_{CH}) + 293.78(n - 1)$$

$$\Delta E^*_{a,n} = -2(E^*_{CC}) + 293.78(n + 1).$$

The (E^*_{CH}) and (E^*_{CC}) quantities (which are necessarily constants in these equations) represent what the individual CH and CC bonds should be if they were truly invariant, in a fictitious situation satisfying the empirically observed constancy of $\Delta E^*_{a,n} - \Delta E^*_{a,n-1}$ by letting $\Delta E^*_{a,n} - \Delta E^*_{a,n-1} = (E^*_{CC}) +$

---

*This observation holds for $n \geqslant 3$, thus including the difference between propane and ethane, and has been verified up to $n = 9$ using the results given in Tables 6.1, 7.1, and 7.2. The differences between ethane and methane are both ~ 2.87 kcal mol^{-1} lower than those indicated above.
†Needless to say, this result is arrived at by taking into consideration all the C atoms of two consecutive normal alkanes and is certainly *not* a representative property of any CH$_2$ group in particular. Indeed, when going from butane to n-pentane, for example, it is meaningless to speculate whether the added group possesses the properties of a butane–CH$_2$ or those of one of the pentane CH$_2$ groups, whose carbon atoms are not identical in the first place. Failure to recognize this point is at the origin of the paradox described in the text and leads to physically unsound views on ring strain in cyclohexane, evaluated on the basis of an assumed noncyclic "model" CH$_2$ group—a problem which is briefly discussed in the next section.

$2(E_{CH}^*)$. This requirement is at variance with the basic non-transferability of constant bond energy terms (Chapter 5.6) and, therefore, physically unacceptable. Similar arguments and conclusions apply also to the (E_{CH}) and (E_{CC}) "bond energies".

So far, we have established that the tentative hypothesis of invariant CC and CH bonds represents a mathematically acceptable, although physically incorrect, way of explaining the constancy of the energy difference between consecutive linear alkanes. The numerical success of popular bond-additivity schemes can thus be accounted for, at least in part. On the other hand, empirical schemes of this type also end up using different bond contributions in order to reproduce particular structural features. In the scheme studied by Laidler[37], for example, the energy terms of primary, secondary, and tertiary CH bonds are not identical, whereas the CC terms are presumed constant. Tatevskii's modification[38] is based on the assumption that not only should the CH bonds in paraffins be classified according to their immediate environment but that the CC bonds should be similarly classified. Here, of course, increasing sophistication and the accompanying proliferation of parameters improve the results but add little to the understanding of the intimate reasons which differentiate the various CC and CH bonds according to their molecular environment. The fact that empirical schemes of this type mimic the correct behavior certainly justifies their utilization but does not bleach the intricacies related to their physical interpretation.

The same distinction between "empirically acceptable" and "physically realistic" solutions is carried over in the following evaluation of vibrational energies, namely, in the interpretation of the constant gain, $\sum^{9n} F(v_i, T) - \sum^{9n-9} F(v_i, T)$, for one added CH_2 group. Following Eqs. 7.9 and 7.11, the quantities

$$\phi_{CH} = \frac{1}{4}(n - 2) \sum^{9n} F(v_i, T) - \frac{1}{4}(n - 1) \sum^{9n-9} F(v_i, T) \qquad (7.13)$$

$$\phi_{CC} = -\frac{1}{2}n \sum^{9n} F(v_i, T) + \frac{1}{2}(n + 1) \sum^{9n-9} F(v_i, T) \qquad (7.14)$$

represent portions of the total vibrational energy formally associated with the individual CH and CC chemical bonds, respectively. In itself, the counting is correct and easy to verify. Indeed, for a $C_n H_{2n+2}$ alkane with $(2n + 2)$ CH and $(n - 1)$ CC bonds, it is $(2n + 2)\phi_{CH} + (n - 1)\phi_{CC} = -\sum^{9n} F(v_i, T)$. The nature of the ϕ_{CH} and ϕ_{CC} terms is also easy to identify, simply by writing Eqs. 7.13 and 7.14 as follows:

$$\sum^{9n} F(v_i, T) = -\left[4\phi_{CH} + \left(\sum^{9n} F(v_i, T) - \sum^{9n-9} F(v_i, T) \right) \right]$$
$$+ n\left[\sum^{9n} F(v_i, T) - \sum^{9n-9} F(v_i, T) \right] \qquad (7.15)$$

$$\sum_{}^{9n} F(v_i, T) = \left[2\phi_{CC} + \left(\sum_{}^{9n} F(v_i, T) - \sum_{}^{9n-9} F(v_i, T) \right) \right]$$

$$+ n \left[\sum_{}^{9n} F(v_i, T) - \sum_{}^{9n-9} F(v_i, T) \right]. \tag{7.16}$$

Because of the invariance of $\sum^{9n} F(v_i, T) - \sum^{9n-9} F(v_i, T)$, each CH_2 group added to the molecule increases its total vibrational energy by a same amount, a fact which is expressed by the linear relationships, Eqs. 7.15 and 7.16. Therefore, both ϕ_{CH} and ϕ_{CC} appear as constants in this type of calculation. For $n = 2$, Eq. 7.15 shows that $\phi_{CH} = -(1/4)[\sum F(v_i, T)]_{CH_4}$ and Eq. 7.16 yields $\phi_{CC} = -[2\phi_{CH} + \sum^{9n} F(v_i, T) - \sum^{9n-9} F(v_i, T)]$.

Incidentally, remembering that $(ZPE + H_T\text{-}H_0)_n = \sum^{9n} F(v_i, T) + 4RT$, Eq. 7.15 takes the form

$$(ZPE + H_T\text{-}H_0)_n = (ZPE + H_T\text{-}H_0)_1 - \left[\sum_{}^{9n} F(v_i, T) - \sum_{}^{9n-9} F(v_i, T) \right]$$

$$+ n \left[\sum_{}^{9n} F(v_i, T) - \sum_{}^{9n-9} F(v_i, T) \right]. \tag{7.17}$$

In this manner Eq. 7.2 is obtained, with a small modification accounting for the empirically observed lowering of $ZPE + E_{vibr}$ accompanying branching. An important corollary (for future use) is

$$n \left[\sum_{}^{9n} F(v_i, T) - \sum_{}^{9n-9} F(v_i, T) \right] - \sum_{}^{9n} F(v_i, T)$$

$$= \sum_{}^{9n} F(v_i, T) - \sum_{}^{9n-9} F(v_i, T) - [F(v_i, T)]_{methane} = constant. \tag{7.18}$$

Let us pause here and examine the results. From the analysis given above it appears that the individual CH and CC bond-energy contributions in C_nH_{2n+2} n-alkanes are most adequately described by Eqs. 7.19 and 7.20.

$$E_{CH} = \varepsilon_{CH}^{av} + \phi_{CH} + \frac{9}{8}RT \tag{7.19}$$

$$E_{CC} = \varepsilon_{CC}^{av} + \phi_{CC} + \frac{9}{4}RT. \tag{7.20}$$

No problem would arise from the use of the correct individual bond energies ε_{ij} at the vibrationless level; it is more practical, however, to consider in each case their appropriate averages over the molecule under study, i.e., $\varepsilon_{CH}^{av} = \Sigma\varepsilon_{CH}/(2n + 2)$ and $\varepsilon_{CC}^{av} = \Sigma\varepsilon_{CC}/(n - 1)$, meaning that E_{CH} and E_{CC} are also averages. Formally, the molecular problem can thus be treated as simply as in the customary two-parameter scheme involving constant CH and CC bond energies. These expressions are certainly "empirically accept-able" because their individual parts add up correctly. The fact alone, how-ever, that the constant contributions ϕ_{CH} and ϕ_{CC} satisfy the constraints

$$-(2\phi_{CH} + \phi_{CC}) = \overset{9n}{\Sigma}F(v_i, T) - \overset{9n-9}{\Sigma}F(v_i, T) = \text{constant} \quad \text{and} \quad (2n+2)\phi_{CH} +$$
$(n-1)\phi_{CC} = -\overset{9n}{\Sigma}F(v_i, T)$ does not imply that it is physically sound to associate constant portions of vibrational energy with the individual chemical bonds. The same observation holds for the RT terms. Briefly, the description of the individual bonds is only approximate, but their sum is exact. Moreover, nonbonded energies being equally included in ΔE_a^* and ΔE_a, Eq. 5.3 is satisfied in an exact manner.

In the description of the energy of a molecule as an additive property of bond contributions, the definition of bond *enthalpy*, H_{ij}, is most convenient for use with thermochemical data and calculations of heats of reaction at, say, 298.15 K. Laidler's scheme[37] is a typical example. Bond enthalpy schemes[19,39,40] are, indeed, popular and simple: from a compilation of average bond enthalpies which are approximately constant in a series of similar compounds, it is easy to obtain enthalpies of atomization, $\Delta H_a = \Sigma H_{ij}$, to a good approximation. When appreciable errors are found, the reason is usually sought in terms of special factors in the molecular structure, such as partial double bond character, ionic character, and steric strains or repulsions. In terms of bond enthalpies, H_{CH} and H_{CC}, defined for noncyclic saturated hydrocarbons, the following relations hold for C_nH_{2n+2}:

$$\Delta E_a = (n-1)E_{CC} + (2n+2)E_{CH}$$

$$\Delta H_a = (n-1)H_{CC} + (2n+2)H_{CH} = \Delta E_a + (3n+1)RT$$

indicating that

$$H_{CH} = E_{CH} + RT; \quad H_{CC} = E_{CC} + RT. \tag{7.21}$$

The E_{CH} and E_{CC} energies are, of course, those given in Eqs. 7.19 and 7.20. This concludes the description and interpretation of bond energies and enthalpies of noncyclic alkanes. Using these compounds as reference models, we can now proceed with an evaluation of the way ring formation affects vibrational energies—a sort of "strain-energy" at the vibrational level.

9. RING STRAIN AND VIBRATIONAL ENERGIES

The physical interpretation and the quantitative evaluation of strain energies in organic cyclic compounds are of considerable interest for the understanding of both physical and chemical properties of these compounds. The estimation of strain energy from thermochemical data implies a comparison between the energy of the molecule under consideration and that of a "model" of the same compound calculated according to some rules. Under these circumstances, the calculated strain energy depends on the model which is adopted for its evaluation and, understandably, different methods of calculation give different results. Clearly, this is an undesirable situation because the evaluation of a true physical property should not be model-

TABLE 7.7. Evaluation of Strain Energies Relative to Cyclohexane (kcal mol^{-1}) from Thermochemical Data

Molecule	ΔH_f°	ZPE + H_T-H_0	ΔE_a	ΔE_a^*	Strain Energy	
					298.15 K	vibrationless
Cyclohexane	-29.50	107.54	1669.85	1760.81	0.0	0.0
Cyclopentane	-18.46	89.0	1385.52	1460.6	6.0	6.7
Cyclobutane	6.4	70.4	1087.36	1146.5	25.9	27.4
Cyclopropane	12.74	51.9	807.73	851.0	27.2	29.4

The "model" ΔE_a (and ΔE_a^*) values with reference to cyclohexane are deduced from (1/6) ΔE_a (cyclohexane) [viz. (1/6) ΔE_a^*] times the number of CH_2 groups in the cycle. Strain energy is defined in each case as the difference between the "model" ΔE_a (or ΔE_a^*) and the corresponding experimental value.

dependent. The problem, therefore, is one of deciding which part, if any, of the strain energy derived from model-dependent thermochemical considerations has a genuine physical origin and, on the other hand, what part is an artefact due to the selected mode of calculation.

At the vibrationless level, theory and extensive comparisons with experiment (Chapter 6) indicate that saturated hydrocarbons constructed from chair and/or boat cyclohexane rings can be treated exactly in the same manner as their noncyclic counterparts, with no change in the parametrization of the appropriate equations describing CC and CH bond energy contributions. No "special effect" which could go under the heading "ring strain" occurs specifically because of the cyclic structure of cyclohexane*. This is no longer true for smaller cycles (Table 7.7): valid comparisons can be made because they concern in each case only CH_2 groups which are necessarily electroneutral, unlike their counterparts in normal alkanes. Ring strain (with respect to cyclohexane) can be qualitatively traced back to the fact that the electron distributions differ significantly from those of cyclohexane, namely as regards marked gains in $2p$ electrons at carbon as the rings become smaller (Table 4.8), not to speak about factors related to changes in internuclear distances.

For molecules at some temperature T, however, an apparent "strain energy" would seem to arise as the result of a systematic error introduced

---

*If the increment $\Delta E_{a,n}^* - \Delta E_{a,n-1}^* = 293.78$ kcal mol^{-1} is tentatively interpreted as a representative contribution of one CH_2 group, the atomization energy ΔE_a^* estimated for cyclohexane, 1762.68 kcal mol^{-1}, would suggest a strain of 1.87 kcal mol^{-1} for this molecule. This reasoning is not appropriate, however. The 293.78 kcal mol^{-1} term includes the change in $\lambda_1 \Sigma N_{CC} \Delta q_C + \lambda_2 \Sigma \Delta q_C$ (computed over all C atoms) in going from one linear paraffin molecule to the next one, which is definitely not a representative property of a CH_2 group. There is no point in deliberately using an erroneous estimate for $\lambda_1 \Sigma N_{CC} \Delta q_C + \lambda_2 \Sigma \Delta q_C$ when calculating cyclohexane because, on that count, any molecule could be easily misrepresented. Briefly, the tentative (and not uncommon) interpretation of the 1.87 kcal mol^{-1} error in terms of an alledged "strain energy" in cyclohexane is a bad case of mistaken identity.

in any comparison of cyclic *vs.* noncyclic structures, as pointed out by Nelander and Sunner[41]. This is a point requiring immediate attention. Of course, knowing that bond energies defined at the vibrationless level give the correct corresponding atomization energies ΔE_a^*, we take the hint and focus attention on the vibrational part.

The exact relationship describing $C_nH_{2n+2-2m}$ saturated hydrocarbons with m rings, decomposing to give $(3n + 2 - 2m)$ atoms, is (Eq. 5.2)

$$\Delta H_a = \Delta E_a + (3n + 1 - 2m)RT.$$

On the other hand, considering the $(n + m - 1)$ CC bonds and the $(2n + 2 - 2m)$ CH bonds of these molecules, it follows from the bond enthalpy scheme (Eq. 7.21) that

$$\Delta H_a^{model} = \Delta E_a^{model} + (3n + 1 - m)RT.$$

In all discussions of strain energy E_{strain}, it is taken to be a property of the isolated molecule and must therefore be defined in terms of *energy* and not of enthalpy. Hence, for the hydrocarbon with m rings, it is given by

$$E_{strain} = \Delta E_a^{model} - \Delta E_a = \Delta H_a^{model} - \Delta H_a - mRT;$$

i.e., defining now $-D = \Delta H_a^{model} - \Delta H_a$ as the error introduced by assuming rigidly that cycloalkanes behave in every respect like their noncyclic counterparts,

$$\Delta E_a^{model} - \Delta E_a = -(D + mRT). \tag{7.22}$$

The exact value of ΔE_a follows from Eq. 5.3, i.e., for $C_nH_{2n+2-2m}$,

$$\Delta E_a = \Delta E_a^* - \sum^{9n-6m} F(v_i, T) + \frac{1}{2}(9n - 6m)RT.$$

The "model" atomization energy deduced from Eqs. 7.19 and 7.20, implying that cyclic and noncyclic saturated hydrocarbons are treated from the outset in precisely the same manner, is

$$\Delta E_a^{model} = \Delta E_a^* + n(\phi_{CC} + 2\phi_{CH}) + (m - 1)(\phi_{CC} - 2\phi_{CH}) + \frac{9}{2}nRT.$$

Comparison with the exact expression for ΔE_a by means of Eq. 7.22 gives

$$\sum^{9n-6m} F(v_i, T) + 4RT = -n(\phi_{CC} + 2\phi_{CH})$$
$$+ (1 - m)(\phi_{CC} - 2\phi_{CH} + 4mRT) - D.$$

Equations 7.13 and 7.14 indicate that $-(\phi_{CC} + 2\phi_{CH}) = \sum^{9n} F(v_i, T) - \sum^{9n-9} F(v_i, T)$ and, using also Eq. 7.18, that $(\phi_{CC} - 2\phi_{CH}) = [\sum F(v_i, T)]_{methane} - [\sum^{9n} F(v_i, T) - \sum^{9n-9} F(v_i, T)]$. Finally, remembering that $ZPE + H_T$-$H_0 = \sum F(v_i, T) + 4RT$, it is found that

$$ZPE + H_T\text{-}H_0 = n\left[\sum^{9n} F(v_i, T) - \sum^{9n-9} F(v_i, T)\right] + (1-m)\left\{(ZPE + H_T\text{-}H_0)_1\right.$$
$$\left. - \left[\sum^{9n} F(v_i, T) - \sum^{9n-9} F(v_i, T)\right]\right\} - D. \tag{7.23}$$

Equation 7.23 is visibly the counterpart of the empirical equation which has been successfully applied earlier in comparisons with experimental data (Section 4, Eq. 7.5). The clue to its derivation is in the empirically observed near-constancy of the difference $\left[\sum^{9n} F(v_i, T) - \sum^{9n-9} F(v_i, T)\right]$ and in the corollary (Eq. 7.18) derived therefrom. This is about as much as one can say about Eq. 7.23 because it would be incorrect to assert that the $ZPE + H_T\text{-}H_0$ energy content of cycloalkanes has been theoretically deduced from that of their noncyclic parent compounds. Indeed, vibrational energies are not truly partitionable quantities, meaning that the bond properties (Eqs. 7.19 and 7.20) can only be regarded as approximations. The error term, D, is there to pick up whatever residual differences exist between expectations based on Eq. 7.23 and experiment. Under these circumstances, it is rewarding to observe that D, to be discussed further below, is relatively small and under control (in that it can be evaluated from empirical rules) showing that the vibrational energy content of cyclic and noncyclic saturated hydrocarbons is, for the largest part by far, a quantity which can be constructed in a simple additive fashion. Rule 1 described in Section 4 is recognized in the coefficient of n which is the same as in Eq. 7.17 and rule 3 expresses the fact that the coefficient of $(1 - m)$ is simply the constant term of Eq. 7.17 describing normal alkanes. Rule 2, expressing an empirically observed lowering (~ 0.3 kcal mol^{-1}) of vibrational energy upon branching, is transferred to cycloalkanes. Numerical analyses also indicate that a correction of 0.296 kcal mol^{-1} is required at 298.15 K for each CC bond of the cycle, but not so at 0 K. It appears reasonable, therefore, to describe this empirical correction as $\sim RT/2$ (rule 4), to be included in D. The temperature dependence of the butane *gauche* correction (Eq. 7.5), however, is presently not established.

It is interesting that the corrections included in D actually lower the predicted vibrational energy content of a cycloalkane relative to its model. Note that a lowering in vibrational energy corresponds to an actual *stabilization* and translates here, roughly speaking, as a loss in flexibility of a cycloalkane relative to its noncyclic counterpart. Similarly, the $ZPE + H_T\text{-}H_0$ energies of cyclobutane (70.4 ± 0.1 kcal mol^{-1}, from spectroscopic data[42,43]) and cyclopropane (51.9 ± 0.3 kcal mol^{-1}, from vibrational analyses[44,45]) are also lower (by 1.6 ± 0.3 and 2.2 ± 0.5 kcal mol^{-1}, respectively) than what would be inferred from cyclohexane (Eq. 7.23) and reflect a relative stabilization at the vibrational level accompanying an increased ridigity of the molecular framework. These examples illustrate in what manner reduced vibrational contributions occasionally make up in part for major destabilization occurring at the vibrationless level.

While describing important aspects regarding bond- and vibrational energies in a spirit of unification, the present analysis should by no means be interpreted as a veiled attempt to render their construction from additivity rules any more theoretical and rigorous than it can possibly be. Vibrational energies are not truly additive or partitionable properties; it is a fortunate circumstance that they can be adequately represented in a simple fashion. In this particular instance, Nature has been generous, indeed, in not imposing too much strain on us, in our efforts to keep the description of molecular energies at a manageable but still realistic level.

10. SUMMARY

Experimental enthalpies of formation and atomization energies of vibrationless molecules deduced from theory enable the evaluation of ZPE + H_T-H_0 energies which, for selected classes of compounds, are in good agreement with their spectroscopic counterparts. This possibility of deriving ZPE + H_T-H_0 energies offers a way of recognizing regularities from an examination of numerous cases. Noteworthy aspects are (i) the decrease in ZPE + H_T-H_0 energy by $\sim RT/2$ for each hindered internal rotation in cyclic compounds, (ii) its invariance with respect to the change from the chair to the boat conformation of the cyclohexane ring, and (iii) the decrease in vibrational energy by ~ 0.85 kcal mol^{-1} for one *gauche* interaction.

Methyl-substituted cyclohexane isomers differing by the conformational position of one methyl group reveal a destabilization of ~ 3.6 kcal mol^{-1} of the axial *vs.* the equatorial position at the vibrationless level, which is partially compensated by a ~ 1.7 kcal mol^{-1} lowering in vibrational energy so that, ultimately, the axial position of a methyl group is only ~ 1.9 kcal mol^{-1} less stable than the equatorial position. In comparisons between isomers, the order of stability indicated by the ΔE_a^* energies is in most cases also reflected in the $\Delta H_f°$'s. Occasionally, however, this order may be reversed by the difference in vibrational energy content between the isomers.

Of course, a number of the results which were described above are well-known and have been applied for quite some time, namely in conformational analysis. The important point, here, is not that they have been deduced but that they were obtained from applications of a theory constructed only upon physical principles which can be readily identified, i.e., ultimately, nuclear-nuclear and nuclear-electronic interaction energies.

REFERENCES

1. S. Fliszár and J.-L. Cantara, *Can. J. Chem.*, **59**, 1381 (1981).
2. R.G. Snyder and J.H. Schachtschneider, *Spectrochim. Acta*, **21**, 169 (1965).
3. F.D. Rossini, "Selected Values of Physical and Thermodynamic Properties of Hydrocarbons and Related Compounds", Carnegie Press, Pittsburgh, PA, 1952.

4. H. Spiesecke and W.G. Schneider, *J. Chem. Phys.*, **35**, 722 (1961).

5. D.M. Grant and E.G. Paul, *J. Am. Chem. Soc.*, **86**, 2984 (1964).

6. L.P. Lindeman and J.Q. Adams, *Anal. Chem.*, **43**, 1245 (1971).

7. D.A. Pittam and G. Pilcher, *J. Chem. Soc., Faraday Trans. I*, **68**, 2221 (1972).

8. G. Pilcher and J.D.M. Chadwick, *Trans. Faraday Soc.*, **63**, 2357 (1967).

9. (a) G.R. Somayajulu and B.J. Zwolinski, *J. Chem. Soc., Faraday Trans. II*, **68**, 1971 (1972); *ibid.*, **70**, 967 (1974); (b) W.D. Good, *J. Chem. Thermodyn.*, **2**, 237 (1970).

10. R.H. Boyd, S.N. Sanwal, S. Chary–Tehany, and D.M. McNally, *J. Phys. Chem.*, **75**, 1264 (1971).

11. H. Henry, G. Kean, and S. Fliszár, *J. Am. Chem. Soc.*, **99**, 5889 (1977).

12. D.K. Dalling and D.M. Grant, *J. Am. Chem. Soc.*, **89**, 6612 (1967).

13. J.B. Stothers, "Carbon–13 NMR Spectroscopy", Academic Press, New York, NY, 1972.

14. D.K. Dalling, D.M. Grant, and E.G. Paul, *J. Am. Chem. Soc.*, **95**, 3718 (1973).

15. D.K. Dalling and D.M. Grant, *J. Am. Chem. Soc.*, **96**, 1827 (1974).

16. G.E. Maciel and H.C. Dorn, *J. Am. Chem. Soc.*, **93**, 1268 (1971).

17. G.E. Maciel, H.C. Dorn, R.L. Green, W.A. Kleschick, M.R. Peterson, and G.H. Wahl, *J. Org. Magn. Res.*, **6**, 178 (1974).

18. H. Beierbeck and J.K. Saunders, *Can. J. Chem.*, **55**, 3161 (1977).

19. J.D. Cox and G. Pilcher, "Thermochemistry of Organic and Organometallic Compounds", Academic Press, New York, NY, 1970.

20. C.J. Egan and W.C. Buss, *J. Phys. Chem.*, **63**, 1837 (1959).

21. J.K. Choi, M.J. Joncich, Y. Lambert, P. Deslongchamps, and S. Fliszár, *J. Mol. Struct. Theochem.* **89**, 115 (1982).

22. D.J. Subach and E. Zwolinski, *J. Chem. Eng. Data*, **20**, 232 (1975).

23. J.P. Huvenne, G. Vergoten, G. Fleury, S. Odiot, and S. Fliszár, *Can. J. Chem.*, **60**, 1347 (1982).

24. N.L. Allinger and M.T. Wuesthoff, *J. Org. Chem.*, **36**, 2051 (1971).

25. (a) W.G. Dauben, O. Rohr, A. Labbauf, and F.D. Rossini, *J. Phys. Chem.*, **64**, 283 (1960); (b) G.F. Davies and E.C. Gilbert, *J. Am. Chem. Soc.*, **63**, 1585 (1971); D.M. Speros and F.D. Rossini, *J. Phys. Chem.*, **64**, 1723 (1960); T. Miyazawa and K.S. Pitzer, *J. Am. Chem. Soc.*, **80**, 60 (1958).

26. B.L. Crawford, J.E. Lancaster, and R.G. Inskeep, *J. Chem. Phys.*, **21**, 678 (1953).

27. J.E. Kilpatrick and K.S. Pitzer, *J. Res. Nat. Bur. Stand.*, **38**, 191 (1948).

28. A.J. Baines and J.D.R. Howells, *J. Chem. Soc., Faraday Trans. II*, **69**, 532 (1973).

29. N. Sheppard, *J. Chem. Phys.*, **17**, 74 (1949).

30. W.C. Harris and I.W. Levin, *J. Chem. Phys.*, **54**, 3227 (1971).

31. W.C. Harris and I.W. Levin, *J. Mol. Spectrosc.*, **39**, 441 (1971).

32. P. Cossee and J.H. Schachtschneider, *J. Chem. Phys.*, **44**, 97 (1966).

33. Z. Buric and P.J. Krueger, *Spectrochim. Acta*, **30A**, 2069 (1974).

34. R.G. Snyder and G. Zerbi, *Spectrochim. Acta*, **23A**, 391 (1967).

35. K. Hamada and H. Morishita, *Z. Physik. Chem.*, **97**, 295 (1975).

36. S. Fliszár, *J. Am. Chem. Soc.*, **102**, 6946 (1980).

37. K.J. Laidler, *Can. J. Chem.*, **34**, 626 (1956); E.G. Lovering and K.J. Laidler, *Can. J. Chem.*, **38**, 2367 (1960).

38. V.M. Tatevskii, V.A. Benderskii, and S.S. Yarovoi, "Rules and Methods for Calculating the Physico-Chemical Properties of Paraffinic Hydrocarbons", B.P. Mullins, Translation Ed., Pergamon Press, Oxford, 1961.

39. L. Pauling, "Nature of the Chemical Bond", Cornell University Press, Ithaca, 1960.

40. S. Benson, *J. Chem. Educ.*, **42**, 502 (1965).

41. B. Nelander and S. Sunner, *J. Chem. Phys.*, **44**, 2476 (1966).

42. F.A. Miller, R.J. Capwell, R.C. Lord, and D.G. Rea, *Spectrochim. Acta*, **28A**, 603 (1972).

43. R.C. Lord and I. Nakagawa, *J. Chem. Phys.*, **39**, 2951 (1963).

44. M. Dupuis and J. Pacanski, *J. Chem. Phys.*, **76**, 2511 (1982).

45. A.W. Baker and R.C. Lord, *J. Chem. Phys.*, **23**, 1636 (1955).

CHAPTER 8

Unsaturated Hydrocarbons

1. INTRODUCTION

The first part of this Chapter deals with simple $R_1R_2C = CR_3R_4$ hydrocarbons, where the R's are alkyl groups or hydrogen. The calculations are carried out without explicit consideration of nonbonded Coulomb interactions. This approximation introduces an error similar to that encountered in the study of saturated hydrocarbons but, because of other uncertainties affecting the evaluation of olefinic hydrocarbons (namely, as regards their vibrational energies), the accuracy achieved in this manner may still be regarded as acceptable.

The physical concepts underlying the calculations presented for the C_nH_{2n} alkenes are the same as those involved in the work on saturated hydrocarbons. While these concepts remain basically simple, this point may now seem somewhat obscured by the tedious calculations involved in expanding $\Sigma_i\Sigma_j a_{ij}\Delta q_i$ in a way suited for numerical applications. Indeed, the derivation of a general formula for the C_nH_{2n} alkenes has to take into account not only the requirements of charge normalization, in order to get rid of the hydrogen net charges (along the lines described in Chapter 6 for saturated hydrocarbons), but also the fact that the charges at the sp^2 carbons vary both at the σ and the π levels. The problem is, thus, significantly more complex than that of the sp^3 systems. The step-by-step approach, which introduces the necessary approximations one at a time as needed, is developed in detail. The final formula[1] includes all the pertinent information, regarding the class of compounds it describes, in the most condensed fashion, thus permitting straightforward numerical energy calculations from the appropriate sp^3 and

sp^2 carbon–13 NMR spectra. Moreover, the final energy formula for the C_nH_{2n} alkenes offers a general insight into the various charge-dependent contributions to the overall thermodynamic stability of ethylenic compounds (e.g., on the relative importance of σ and π charges in stabilizing the carbon-carbon double bond), including a brief conclusion concerning the relative stabilities of endo- vs. exocyclic double bonds.

The last section offers a rough but simple theoretical treatment of other unsaturated systems, e.g., benzene, dienes, acetylenic compounds, etc., which, far from being complete, hopefully paves the way to future investigations in this area.

2. ENERGY FORMULA FOR C_nH_{2n} OLEFINS

2.1. *Calculation of the a_{ij} Parameters*

In evaluating the a_{ij}'s from Eq. 5.47, we can safely neglect the second-order derivatives because the Δq_i's at sp^2 (and sp^3) carbon atoms turn out to be sufficiently small. The derivatives $(\partial E_C/\partial N)^\circ_\sigma = -0.375$ and $(\partial E_C/\partial N)^\circ_\pi = -0.244$ a.u. for σ and π electrons, respectively, used in these estimates were derived from LCAO–Xα calculations on ethylene[1].

For the CH and CC bonds involving sp^3 carbon atoms, the average inverse distance between nucleus i and the center of electronic charge on atom j (i.e., $\langle r_{ij}^{-1}\rangle^\circ$) is approximated in the usual manner by the inverse of the internuclear distance, $r_{CH} = 1.08$ and $r_{CC} = 1.53$ Å. This approximation locates the centroids of σ charges at the nuclear positions. For sp^2 carbons, however, we consider that the center of π atomic charge is shifted by ~ 0.04 Å from the center of σ charge toward the other carbon engaged in the double bond. The corresponding $\langle r_{ij}^{-1}\rangle^\circ$ parameters [namely, the increase of r_{CH} and r_{CC} (sp^2–sp^3) by ~ 0.02 Å] reflect this hypothesis which is justified *a posteriori* (Section 4) by the results obtained in this fashion. While certainly in qualitative accord with what one would expect *a priori*, the quantitative assessment of the inward shift of the center of $2p$ atomic charge engaged in a bonding π orbital is the product of a numerical analysis revealing that any value other than ~ 0.04 Å would lead to serious systematic errors. The appropriate a_{ij}'s calculated under these conditions are indicated in Table 5.7 ($C\alpha = sp^2$ carbon).

Special attention must be given when atomic electron populations vary at different energy levels. This is the case for vinyl carbons with charges varying at the σ and π levels, i.e.,

$$\Delta q_i = \Delta q_{i\sigma} + \Delta q_{i\pi}.$$

In this case, the $a_{ij}\Delta q_i$ term becomes

$$a_{ij}\Delta q_i = a_{ij}^\sigma \Delta q_{i\sigma} + a_{ij}^\pi \Delta q_{i\pi}$$

indicating that a_{ij} represents now an appropriately weighted average of a_{ij}^σ and a_{ij}^π. As explained in Chapter 5.8, we can take advantage of the relationship $\Delta q_\sigma = m\Delta q_\pi (-1 < m < 0)$ describing the inverse variations in σ and π charges at the sp^2 carbon atoms under study and use Eq. 5.50, i.e.,

$$a_{ij} = \frac{ma_{ij}^\sigma + a_{ij}^\pi}{m+1}. \tag{5.50}$$

The value of m, however, cannot be derived from Mulliken population analyses because calculated σ charges appear to be significantly more dependent on basis set than π charges. The forthcoming numerical analyses suggest that $m \simeq -0.955$.

So far we have collected the theoretical information required for using the basic equations 5.46 and 5.48. Now we can look at the olefins and inquire about the appropriate charges for use in these energy formulas.

2.2. Charge Normalization

The ethylene-C net charge, $q_{C\alpha}^\circ$, is best evaluated using tetramethyl-ethylene as a starter. Its methyl-C NMR shift is 14.3 ppm from ethane, giving (Eq. 4.10) $\Delta q_C = -0.148\delta_C = -2.12$ me relative to the ethane carbon atom; i.e., $q_C = 35.1 - 2.12 = 32.98$ me. From the correlation (described in Chapter 4.6) between q_H and q_C in methyl groups, q_H is estimated at -12.29 me. The net charge of the methyl group is, therefore, -3.89 me and that of the sp^2 carbon is 7.77 me. Assuming $m = -0.955$ and $d\delta_{C\alpha}/dq_\pi \simeq 300$ ppm/e, as explained further below, we find $\Delta q_{C\alpha} \simeq 0.15\delta_{C\alpha}$ me. Tetramethyl-ethylene (δ 123.2 ppm from TMS) and ethylene (δ 122.8 ppm from TMS) differ only by ~ 0.4 ppm, so that the difference in net charge is ~ 0.06 me, ethylene being thus only marginally electron richer than tetramethylethylene. The final estimate is $q_{C\alpha}^\circ \simeq 7.7$ me and, of course, $q_H \simeq -3.85$ me.

The total $(\sigma + \pi)$ net charge $q_{C\alpha}^\circ$ of the ethylene-C atom (relative to 6 e) is also its net σ charge (relative to 5 e) because its net π charge (relative to 1 e) is 0. Hence, a vinyl carbon whose net σ and π charges are q_σ and q_π, respectively, differs from the ethane carbon by $\Delta q_C = q_\sigma + q_\pi - q_C^\circ = (q_\sigma - q_{C\alpha}^\circ) + (q_{C\alpha}^\circ - q_C^\circ) + (q_\pi - 0)$, i.e.,

$$\Delta q_C = \Delta q_\sigma + \Delta q_\pi + \Delta q_{C\alpha}^\circ \tag{8.1}$$

where

$$\Delta q_{C\alpha}^\circ = q_{C\alpha}^\circ - q_C^\circ = -27.4 \text{ me} \tag{8.2}$$

is the difference in net charge between the ethylene and the ethane carbon atoms. The ethylene net charge $(q_{C\alpha}^\circ)$, of course, is used in the evaluation of C=C double bonds, i.e.,

$$\varepsilon_{C=C} = \varepsilon_{C=C}^\circ + a_{C=C}\sum \Delta q_{C\alpha}$$

where $\Delta q_{C\alpha} = q_{C\alpha} - q_{C\alpha}^\circ$ refers to sp^2 carbons, relative to ethylene. The CH and CC simple bonds, however, including those formed by the sp^2 carbon atoms (identified as α-carbons), are expressed with reference to the ethane CH and CC bond energies. This implies that the charge increments $\Delta q_H =$

$q_H - q_H^\circ$ and $\Delta q_C = q_C - q_C^\circ$ are also to be taken relative to the ethane net charges, i.e., $q_C^\circ = 35.1$ me for carbon and $q_H^\circ = -11.70$ me for hydrogen.

The hydrogen net charges appear in the final result (from Eq. 5.48) as the sum $\Sigma \Delta q_H$. The calculation of $\Sigma \Delta q_H = \Sigma q_H - 2n q_H^\circ$ is straightforward. From charge normalization, i.e., $\Sigma q_H = -(\Sigma q_{C\alpha} + \Sigma q_{C\neq\alpha})$, it follows that

$$\Delta q_H = -\left(\sum \Delta q_{C\neq\alpha} + \sum \Delta q_{C\alpha}\right) - 2\Delta q_{C\alpha}^\circ + n q_H^\circ \tag{8.3}$$

where n is the number of C atoms in the C_nH_{2n} alkene. In this fashion, it becomes possible to express ΔE_a^* (Eq. 5.48) by means of a general formula requiring only the charges of the carbon atoms.

2.3. General Formula for C_nH_{2n} Alkenes

The sum of the $\Sigma_i \Sigma_j a_{ij} \Delta q_i$ terms (Eq. 5.48) referring to CC bonds gives

$$\overset{\text{CC bonds}}{\sum_i \sum_j} a_{ij} \Delta q_i = a_{CC} \sum N_{CC} \Delta q_{C\neq\alpha} + a_{C=C} \sum \Delta q_{C\alpha} + a_{C\alpha C} \sum N_{C\alpha C} \Delta q_C.$$

Similarly, we find for the CH bonds that

$$\overset{\text{CH bonds}}{\sum_i \sum_j} a_{ij} \Delta q_i = 4a_{CH} \sum \Delta q_{C\neq\alpha} - a_{CH} \sum N_{CC} \Delta q_{C\neq\alpha} + 2a_{C\alpha H} \sum \Delta q_C$$

$$- a_{C\alpha H} \sum N_{C\alpha C} \Delta q_C + a_{HC} \sum \Delta q_H.$$

A distinction is made between CC and CH bonds involving sp^3 carbons ($\neq\alpha$, described by a_{CC} and a_{CH}) and bonds formed by sp^2 carbons (α-C, described by $a_{C\alpha C}$ and $a_{C\alpha H}$, respectively). Similarly, $N_{C\alpha C}$ is the number of CC bonds formed by α carbons with sp^3 C atoms and N_{CC} is the number CC bonds formed by sp^3 C atoms. Using now Eq. 8.3, the final sum over all the CC and CH bonds becomes

$$\sum_i \sum_j a_{ij} \Delta q_i = \lambda_1 \sum N_{CC} \Delta q_{C\neq\alpha} + \lambda_2 \sum \Delta q_{C\neq\alpha} + n a_{HC} q_H^\circ + (a_{C=C} - a_{HC}) \sum \Delta q_{C\alpha}$$

$$+ (a_{C\alpha C} - a_{C\alpha H}) \overset{\alpha}{\sum} N_{C\alpha C} \Delta q_C + 2a_{C\alpha H} \overset{\alpha}{\sum} \Delta q_C - 2a_{HC} \Delta q_{C\alpha}^\circ \tag{8.4}$$

where

$$\lambda_1 = a_{CC} - a_{CH} = -0.383 \text{ a.u.}$$

$$\lambda_2 = 4a_{CH} - a_{HC} = -0.569 \text{ a.u.}$$

The $\lambda_1 \Sigma N_{CC} \Delta q_{C\neq\alpha} + \lambda_2 \Sigma \Delta q_{C\neq\alpha} + n a_{HC} q_H^\circ$ part of Eq. 8.4 is well-known from the study of saturated hydrocarbons. By means of the charge–^{13}C NMR shift correlation $\Delta q_C = -0.148\delta_C$ me (Eq. 4.10) we can write this part in terms of NMR shifts relative to ethane (taken at ~ 5.8 ppm from TMS), i.e., (with 1 a.u. $= 627.51$ kcal mol^{-1}), $0.0356\Sigma N_{CC}\delta_C + 0.0529\Sigma\delta_C + 7.393n$ kcal mol^{-1}. The term in $\Sigma \Delta q_{C\alpha}$ creates no problem because the charges are expressed with respect to ethylene and can be easily resolved. The terms in

Δq_C, however, which refer to the charges of vinyl carbons with respect to that of ethane, require special attention.

Let us first consider the term $F = (a_{C\alpha C} - a_{C\alpha H})\Sigma^\alpha N_{C\alpha C}\Delta q_C$ of Eq. 8.4. The difference $a_{C\alpha C} - a_{C\alpha H}$ is λ_1^q or λ_1^π, depending on whether it relates to σ or to π charge variations at the vinyl carbons. Of course, λ_1^q is simply the λ_1 value indicated above and $\lambda_1^\pi = a_{C\alpha C}^\pi - a_{C\alpha H}^\pi = -0.379$ a.u. is deduced from the a_{ij}'s given earlier. Instead of using net charges, $q = Z - N$, we begin this calculation in terms of numbers of electrons, remembering that $\Delta N = -\Delta q$. The $\sigma-\pi$ separation gives

$$F = -\lambda_1^\sigma \sum^\alpha N_{C\alpha C}(N_{\sigma v} - N_{\sigma e}) - \lambda_1^\pi \sum^\alpha N_{C\alpha C}(N_{\pi v} - N_{\pi e})$$

where the subscripts v and e stand for vinyl and ethane carbon, respectively. The following transformations are self-explanatory, remembering that σ net charges are expressed with respect to 5, *viz.* 6 e for vinyl and sp^3 carbons, respectively, while π net charges are expressed relative to 1 electron.

$$F = -\lambda_1^\sigma \sum^\alpha N_{C\alpha C}(N_{\sigma v} - 5 - 1 + 6 - N_{\sigma e}) - \lambda_1^\pi \sum^\alpha N_{C\alpha C}(N_{\pi v} - 1 + 1 - 0)$$

$$= -\lambda_1^\sigma \sum^\alpha N_{C\alpha C}(-q_\sigma - 1 + q_C^\circ) - \lambda_1^\pi \sum^\alpha N_{C\alpha C}(-q_\pi + 1)$$

$$= -\lambda_1^\sigma \sum^\alpha N_{C\alpha C}(-q_\sigma + q_{C\alpha}^\circ - q_{C\alpha}^\circ + q_C^\circ - 1) - \lambda_1^\pi \sum^\alpha N_{C\alpha C}(-q_\pi + 1)$$

$$= -\lambda_1^\sigma \sum^\alpha N_{C\alpha C}(-\Delta q_\sigma - \Delta q_{C\alpha}^\circ - 1) - \lambda_1^\pi \sum^\alpha N_{C\alpha C}(-\Delta q_\pi + 1)$$

$$= \lambda_1^\sigma \sum^\alpha N_{C\alpha C}\Delta q_\sigma + \lambda_1^\pi \sum^\alpha N_{C\alpha C}\Delta q_\pi + (\lambda_1^\sigma \Delta q_{C\alpha}^\circ + \lambda_1^\sigma - \lambda_1^\pi) \sum^\alpha N_{C\alpha C}$$

The first two terms of this last expression combine to give $\lambda_{1\alpha}\Sigma N_{C\alpha C}\Delta q_{C\alpha}$, with $\lambda_{1\alpha}$ defined as follows, from Eq. 5.50,

$$\lambda_{1\alpha} = \frac{m\lambda_1^\sigma + \lambda_1^\pi}{m + 1}. \tag{8.5}$$

On the other hand, using the appropriate a_{ij}'s, it is found that

$$(\lambda_1^\sigma \Delta q_{C\alpha}^\circ + \lambda_1^\sigma - \lambda_1^\pi) \sum^\alpha N_{C\alpha C} = 4.19 \sum^\alpha N_{C\alpha C} \text{ kcal mol}^{-1}.$$

With these results, we can now write Eq. 8.4 as follows, in kcal mol^{-1},

$$\sum_i \sum_j a_{ij}\Delta q_i = 0.0356\sum N_{CC}\delta_C + 0.0529\sum\delta_C + 7.393n$$

$$+ (a_{C=C} - a_{HC})\sum\Delta q_{C\alpha} + \lambda_{1\alpha}\sum^\alpha N_{C\alpha C}\Delta q_{C\alpha} \tag{8.6}$$

$$+ 4.19\sum^\alpha N_{C\alpha C} + 2a_{C\alpha H}\sum^\alpha \Delta q_C - 2a_{HC}\Delta q_{C\alpha}^\circ.$$

The calculation of $(a_{C\alpha C} - a_{C\alpha H})\Sigma^\alpha N_{C\alpha C}\Delta q_C$ has presented no difficulty because λ_1^q and λ_1^π are constants, depending only upon molecular geometry and the location of the centers of charge. The important point is that the derivatives $(\partial E_i^{vs}/\partial N_i)^\circ$ which are part of the a_{ij}'s (Eq. 5.47) cancel exactly in λ_1^q and λ_1^π. This is no longer the case with the $2a_{C\alpha H}\Sigma^\alpha \Delta q_C$ term of Eq. 8.6.

The $\sigma-\pi$ separation of $a_{C\alpha H}\Sigma^\alpha\Delta q_C$ gives, by means of Eq. 8.1,

$$a_{C\alpha H}^\sigma \sum^\alpha \Delta q_\sigma + a_{C\alpha H}^\pi \sum^\alpha \Delta q_\pi + 2\langle a_{C\alpha H}\rangle \Delta q_{C\alpha}^\circ.$$

The first two terms are now well-defined with reference to ethylene, and enter the final sum as $2a_{C\alpha H}^\sigma \Sigma^\alpha \Delta q_\sigma + 2a_{C\alpha H}^\pi \Sigma^\alpha \Delta q_\pi$. The $a_{C\alpha H}^\sigma$ and $a_{C\alpha H}^\pi$ values can be calculated from Eq. 5.47, as indicated earlier. The $\langle a_{C\alpha H}\rangle$ parameter, however, represented here loosely as an appropriately chosen average, remains to be clarified. On the physical side, $\langle a_{C\alpha H}\rangle \Delta q_{C\alpha}^\circ$ appears to describe the gain in stability of a CH bond in going from ethane to ethylene due to (i) an increase in electron population of 27.4 me and (ii) the shift of the center of charge (by ~ 0.04 Å along the C=C axis) for 1 electron, which destabilizes the CH bond by $\sim (3/7)0.529[(1/1.08) - (1/1.10)]$ a.u. $\simeq$ 2.395 kcal mol^{-1}. As regards the gain in electronic charge in going from ethane to ethylene, the appropriate a_{CH} values are (from Table 5.7) -0.394 a.u. for ethane and -0.335 a.u. for ethylene. Tentatively taking the average of these values, we obtain in this approximation that

$$\langle a_{C\alpha H}\rangle \Delta q_{C\alpha}^\circ = \left[\frac{1}{2}(a_{CH} + a_{C\alpha H}^\sigma)\right]\Delta q_{C\alpha}^\circ - \frac{3}{7}0.529\left(\frac{1}{1.08} - \frac{1}{1.10}\right) \text{a.u.}$$

$$= 3.872 \text{ kcal mol}^{-1}.$$

The validity of this estimate can be evaluated as follows. The 3.872 term represents the gain in stability of a CH bond due to the change $C(sp^3) \rightarrow C(sp^2)$ from ethane to ethylene. Leaving the ethane-H charge as reference for the $a_{HC}\Delta q_H$ part, the new $\varepsilon_{CH}^\circ(sp^2\text{—H})$ reference energy becomes $106.806 + 3.872 = 110.68$ kcal mol^{-1}. Using now the ethylene-H net charge, -3.85 me (corresponding to its C charge of 7.7 me), i.e., $\Delta q_H = 7.85$ me, it results that $a_{HC}\Delta q_H = -4.96$ and $\varepsilon_{CH} = 105.72$ kcal mol^{-1}. Four of these CH bonds plus the theoretical $\varepsilon_{C=C}^\circ = 139.27$ double bond contribution lead to ΔE_a^* (ethylene) $= 562.15$ kcal mol^{-1}, which compares favorably with its experimental counterpart, 562.22 kcal mol^{-1}. Similarly, remembering that the gain for a $C(sp^2)$—$C(sp^3)$ bond is 4.19 kcal mol^{-1} larger than that for a $C(sp^2)$—H bond, i.e., $\langle a_{C\alpha C}\rangle \Delta q_{C\alpha}^\circ = \langle a_{C\alpha H}\rangle \Delta q_{C\alpha}^\circ + 4.19$ kcal mol^{-1}, we can define a new $\varepsilon_{CC}^\circ(sp^2\text{—}sp^3)$ reference as $69.633 + 3.872 + 4.19 = 77.70$ kcal mol^{-1}, where the Δq_C's are taken relative to ethylene and ethane for the sp^2 and sp^3 carbons, respectively.

The $\langle a_{C\alpha H}\rangle \Delta q_{C\alpha}^\circ$ term appears twice for $a_{C\alpha H}\Sigma^\alpha\Delta q_C$ which, in turn, enters into Eq. 8.6 with a factor of 2. Hence, taking now also into account the $-2a_{HC}\Delta q_{C\alpha}^\circ$ term of Eq. 8.6, the constant term appearing in this equation is

$$4\langle a_{C\alpha H}\rangle \Delta q_{C\alpha}^\circ - 2a_{HC}\Delta q_{C\alpha}^\circ = -19.14 \text{ kcal mol}^{-1}. \tag{8.8}$$

In comparisons with experimental atomization energies deduced from Eq. 5.5, this result, combined with the theoretical $\varepsilon_{C=C}^\circ = 139.27$ kcal mol^{-1}

energy, leads to a systematic deviation of ~ 0.08 kcal mol^{-1}. Of course, it is not possible to decide which of Eq. 8.8 or $\varepsilon^\circ_{C=C}$ is "wrong" because each term appears once for a $C=C$ double bond system. At this level of accuracy, however, this question is academic. Tentatively assuming the $\varepsilon^\circ_{C=C}$ value to be correct, we modify the result indicated in Eq. 8.8 from -19.14 to -19.06 kcal mol^{-1}. Incidentally, it is noteworthy that the location of the π center of charge which has been postulated affects both the result given in Eq. 8.8 and the 4.19 kcal mol^{-1} coefficient of $\Sigma^\alpha N_{C\alpha C}$. Moreover, it enters the calculation of $a_{C=C}$. This ~ 0.04 Å parameter satisfies thus 3 numerical values which play important roles in the final energy formula.

This analysis of the $a^\sigma_{C\alpha H}\Sigma^\alpha \Delta q_C$ term still leaves us with $2a^\sigma_{C\alpha H}\Sigma^\alpha \Delta q_\sigma + 2a^\pi_{C\alpha H}\Sigma^\alpha \Delta q_\pi$ which are part of Eq. 8.6. Combining now these terms with $(a_{C=C} - a_{HC})\Sigma^\alpha \Delta q_{C\alpha}$, which is easily separated to give $(a^\sigma_{C=C} - a_{HC})\Sigma^\alpha \Delta q_\sigma + (a^\pi_{C=C} - a_{HC})\Sigma^\alpha \Delta q_\pi$, we define

$$\lambda^\sigma_3 = a^\sigma_{C=C} + 2a^\sigma_{C\alpha H} - a_{HC} = -0.465 \text{ a.u.}$$

$$\lambda^\pi_3 = a^\pi_{C=C} + 2a^\pi_{C\alpha H} - a_{HC} = -0.347 \text{ a.u.}$$

and obtain from Eq. 5.50 that

$$\lambda_3 = \frac{m\lambda^\sigma_3 + \lambda^\pi_3}{m + 1} \tag{8.9}$$

In this manner, Eq. 8.6 reduces to

$$\sum_i \sum_j a_{ij}\Delta q_i = 0.0356 \sum N_{CC}\delta_C + 0.0529 \sum \delta_C + \lambda_{1\alpha}\sum^\alpha N_{C\alpha C}\Delta q_{C\alpha}$$
$$+ \lambda_3 \sum \Delta q_{C\alpha} + 4.19 \sum^\alpha N_{C\alpha C} + 7.393n - 19.06 \text{ kcal mol}^{-1} \tag{8.10}$$

The important point is that the $\Delta q_{C\alpha}$'s are expressed with respect to the proper reference, which is ethylene, in terms of increments in total $(\sigma + \pi)$ net charge. Moreover, the inverse variations of σ and π charges are now also properly taken into account in Eq. 8.10.

The final step consists in translating this formula into an equivalent one in which the $\Delta q_{C\alpha}$'s are expressed in terms of carbon–13 nuclear magnetic resonance shifts. The latter are, of course, to be expressed relative to ethylene, which is taken at δ 122.8 ppm from tetramethylsilane. The results presented in Chapter 4.3 and, namely, Fig. 4.3, indicate that this is possible. Using the relationships $\Delta q_\sigma = m\Delta q_\pi$ and $\Delta q_{C\alpha} = \Delta q_\sigma + \Delta q_\pi = (m + 1)\Delta q_\pi$, it follows that

$$\Delta q_{C\alpha} = \frac{m + 1}{d\delta_{C\alpha}/dq_\pi}\delta_{C\alpha}$$

where $d\delta_{C\alpha}/dq_\pi$ is the derivative of $\delta_{C\alpha}$ with respect to π populations. The advantage of this transformation is twofold. Firstly, we can eliminate $m + 1$ from the final expression, which is desirable because $m + 1$ is small and, thus, likely to introduce a significant error owing to the fact that m itself is not well-known. Secondly, it appears somewhat easier to obtain a realistic estimate for $d\delta_{Ca}/dq_\pi$ than for a derivative with respect to total charge, because σ charges are particularly difficult to extract from population analyses. It is now possible to express λ_{1a} (Eq. 8.5) and λ_3 (Eq. 8.9) in kcal mol^{-1} ppm^{-1} units, using also the conversion factor 1 a.u. = 627.51 kcal mol^{-1}. In this manner we obtain

$$\lambda_{1a} = \frac{(m\lambda_1^\sigma + \lambda_1^\pi)627.51}{d\delta_{Ca}/dq_\pi} \text{ kcal mol}^{-1} \text{ ppm}^{-1}$$

$$\lambda_3 = \frac{(m\lambda_3^\sigma + \lambda_3^\pi)627.51}{d\delta_{Ca}/dq_\pi} \text{ kcal mol}^{-1} \text{ ppm}^{-1}$$

where $\lambda_1^\sigma, \lambda_1^\pi, \lambda_3^\sigma$ and λ_3^π are in atomic units and the derivative is in ppm/electron. Within the limits of validity of the charge–shift relationship, Eq. 8.10 becomes

$$\sum_i \sum_j a_{ij} \Delta q_i = 0.0356 \sum N_{CC} \delta_C + 0.0529 \sum \delta_C + \lambda_{1a} \sum N_{CaC} \delta_{Ca} + \lambda_3 \sum \delta_{Ca}$$
$$+ 4.19 \sum N_{CaC} + 7.393n - 19.06 \text{ kcal mol}^{-1}. \qquad (8.11)$$

The real difficulty rests with the "true" values of m and $d\delta_{Ca}/dq_\pi$ because atomic charges calculated from Mulliken's scheme are strongly basis set dependent, particularly as regards σ charges. Charge separations are invariably larger when increasingly larger basis sets are used in *ab initio* calculations, to the point that unrealistic results are obtained. Charge analyses carried out at the STO–3G level (Chapter 4.3) suggest a value of ~ 250 ppm/e for the $d\delta_{Ca}/dq_\pi$ slope, which is probably also a result of slightly overestimated π charges. In selecting tentatively $d\delta_{Ca}/dq_\pi \simeq 300$ ppm/e for vinyl C atoms, we regard this value only as a reasonable estimate but cannot attach any precision to it. This holds also for the $m \simeq -0.955$ value which is tentatively selected. With these premises, we deduce $\lambda_{1a} \simeq -0.028$ and $\lambda_3 \simeq 0.20$ kcal mol^{-1} ppm^{-1}, but it is clear that similar values can also be obtained from alternate choices for m and $d\delta_{Ca}/dq_\pi$. The important point is that λ_{1a} is small and negative, whereas λ_3 is relatively large and positive. The approximate validity of our rough estimates for m and $d\delta_{Ca}/dq_\pi$ is best illustrated by comparisons with experimental results. Indeed, using $\lambda_{1a} = -0.028$ and $\lambda_3 = 0.20$ kcal mol^{-1} ppm^{-1} in applications of Eq. 8.11, it turns out that ethylene, 1-alkenes, *trans*-alkenes and tetramethylethylene are calculated with an average error of ~ 0.2 kcal mol^{-1}. Numerical applications indicate that the following simplified expression (Eq. 8.12), in which $\lambda_{1a} = 0$, gives equally satisfactory results, within the limits of uncertainties set by the precision of the experimental data. This simplification seems reasonable, considering the uncertainties associated with λ_{1a} and λ_3.

$$\sum_i \sum_j a_{ij} \Delta q_i = 0.0356 \sum N_{CC} \delta_{C \neq \alpha} + 0.0529 \sum \delta_{C \neq \alpha} + 0.18 \sum \delta_{Ca}$$
$$+ 7.393n + 4.19 \sum N_{CaC} - 19.06 \text{ kcal mol}^{-1}. \qquad (8.12)$$

This is the basic formula which is used in the forthcoming comparisons with experimental results.

3. NUMERICAL APPLICATIONS: C_nH_{2n} OLEFINS

The atomization energies of olefins can now be readily deduced from their ^{13}C NMR spectra[2], using Eqs. 5.48 and 8.12. The calculation of the corresponding enthalpies of formation, however, also requires a realistic estimate of zero-point and heat-content energies, i.e., of ZPE + H_T-H_0, for use in Eq. 5.5. Spectroscopic information is unfortunately scarce for the ethylenic compounds under study (Chapter 7.5).

Fortunately, however, we can profit from a detailed study on a representative variety of saturated hydrocarbons indicating that one CH_2 added to an alkyl group increases ZPE + H_T-H_0 by 18.213 kcal mol^{-1} whereas branching lowers this energy by 0.343 kcal mol^{-1} (Chapter 7.3). (Situations of extreme steric crowding, such as those arising with the presence of two *tert*-butyl groups attached to a same carbon atom are not represented by these rules.) It seems reasonable to assume a similar pattern for the normal and branched ethylenic hydrocarbons. This approximation is expressed by the following equation:

$$\text{ZPE} + H_T\text{-}H_0 \simeq 33.35 + 18.213(n - 2) - 0.343\, n_{\text{branching}}. \qquad (7.6)$$

The experimental results given in Chapter 7.5 certainly support this approximation although, unfortunately, it contains no information regarding a number of important questions, e.g., about a possible effect accompanying a change from *trans* to *cis* geometry. It is noteworthy, in any case, that numerous enthalpies of formation calculated from the appropriate ΔE_a^* atomization energies (Eqs. 5.48 and 8.12) and Eq. 5.5 are in satisfactory agreement with their experimental counterparts, as indicated in Table 8.1.

The results deduced from Eq. 8.12 concern only monoalkyl- and *trans*-dialkylsubstituted ethylenes, and tetramethylethylene. They agree within ~0.2 kcal mol^{-1} (average deviation) with their experimental counterparts. Before proceeding with the study of *cis*- and *gem*-disubstituted and trialkyl-substituted ethylenes, it becomes important to examine the geometries of the compounds investigated here. *Ab initio* calculations[5] indicate the results given below.

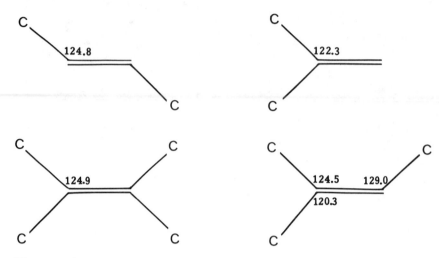

These results correspond well to what one would anticipate in the first place and it appears reasonable to assume that the hybridization of the sp^2 carbons is the same (or nearly so) in all the compounds described by Eq. 8.12. Indeed, a change in local geometry about the vinyl carbons would cause an additional difficulty linked to the σ–π charge separation, affecting the relationship $q_\sigma = mq_\pi + \text{constant}$. This point must be considered in the evaluation of *cis* olefins. If we write $q_\sigma^{cis} = mq_\pi^{cis} + c^{cis}$ for the latter and describe their Δq_σ and Δq_π charge increments with respect to reference charges taken from the relationship $q_\sigma^{trans} = mq_\pi^{trans} + c^{trans}$ (which describes *trans*- or 1-olefins), a nonzero difference $c^{cis} - c^{trans}$ introduces a modification in the coefficient of $\Sigma N_{C\alpha C}$ and in the constant term of Eq. 8.12. Small effects of this sort resist theoretical evaluation because $c^{cis} - c^{trans}$ cannot be determined in a reliable manner. On the other hand, an empirical assessment is presently also difficult because it is not known how accurately Eq. 7.6 approximates the ZPE + H_T-H_0 values of *cis* olefins. It appears, however, that the approximation

$$\sum_i \sum_j a_{ij}\Delta q_i \simeq 0.0356 \sum N_{CC}\delta_{C\neq\alpha} + 0.0529 \sum \delta_{C\neq\alpha} + 0.18 \sum \delta_{C\alpha}$$
$$+ 7.393n + 4.0 \sum N_{C\alpha C} - 18.12 \text{ kcal mol}^{-1} \tag{8.13}$$

combined with Eqs. 7.6 and 5.5 yields satisfactory standard enthalpies of formation for *cis*-olefins. The results for trisubstituted ethylenes are deduced in the same manner (Table 8.1). The overall average deviation (0.30 kcal mol^{-1}) is certainly acceptable and it appears fair to conclude that, in spite of some difficulties and uncertainties, the charge dependence of the bond energy contributions which is predicted by theory has been verified at a very demanding level, that of experimental accuracy.

It remains that a good numerical agreement between theory and experiment, though obviously welcome, is not necessarily always sufficient to satisfy our minds. The physical concepts underlying the calculations pre-

TABLE 8.1. Comparison Between Calculated and Experimental Enthalpies of Formation of Unsaturated Hydrocarbons (kcal mol^{-1})

Molecule	$\Sigma \varepsilon_{ij}^\circ$	$\Sigma_i \Sigma_j a_{ij} \Delta q_i$	ΔE_a^*	ΔH_f° (298.15, gas)		
				Calcd.	Benson[4]	Exptl.
Ethene	566.49	−4.27	562.22	12.38	12.22	12.50
Propene	849.74	8.90	858.64	4.80	4.78	4.88
1-Butene	1132.98	19.57	1152.55	−0.27	−0.01	−0.03
(Z)2-Butene	1132.98	20.84	1153.83	−1.54	−1.97	−1.67
(E)2-Butene	1132.98	22.01	1154.99	−2.71	−2.97	−2.67
2-Me-Propene	1132.98	23.52	1156.51	−4.22	−3.56	−4.04
1-Pentene	1416.23	30.18	1446.41	−5.28	−4.96	−5.00
(Z)2-Pentene	1416.23	31.65	1447.88	−6.75	−6.73	−6.71
(E)2-Pentene	1416.23	32.49	1448.72	−7.60	−7.73	−7.59
2-Me-1-Butene	1416.23	33.94	1450.17	−9.05	−8.32	−8.68
3-Me-1-Butene	1416.23	31.43	1447.66	−6.87	−6.78	−6.92
2-Me-2-Butene	1416.23	35.08	1451.31	−10.19	−10.30	−10.17
1-Hexene	1699.47	40.28	1739.76	−9.80	−9.88	−9.96
(Z)2-Hexene	1699.47	42.12	1741.60	−11.64	−11.68	−12.51
(E)2-Hexene	1699.47	43.09	1742.57	−12.61	−12.68	−12.88
(Z)3-Hexene	1699.47	42.38	1741.86	−11.90	−11.49	−11.38
(E)3-Hexene	1699.47	43.10	1742.58	−12.62	−12.49	−13.01
2-Me-1-Pentene	1699.47	44.35	1743.82	−13.86	−13.27	−14.19
3-Me-1-Pentene	1699.47	41.75	1741.22	−11.60	−11.73	−11.82
4-Me-1-Pentene	1699.47	42.04	1741.52	−11.90	−11.17	−12.24
(Z)3-Me-2-Pentene	1699.47	45.81	1745.29	−15.33	−15.06	−15.08
(E)3-Me-2-Pentene	1699.47	45.24	1744.71	−14.75	−15.06	−14.86
(Z)4-Me-2-Pentene	1699.47	43.69	1743.16	−13.54	−15.53	−13.73
(E)4-Me-2-Pentene	1699.47	44.53	1744.00	−14.38	−14.53	−14.69
2-Et-1-Butene	1699.47	43.44	1742.91	−12.95	−13.18	−13.38
2,3-diMe-1-Butene	1699.47	45.47	1744.95	−15.33	−14.65	−15.85
3,3-diMe-1-Butene	1699.47	43.41	1742.89	−13.61	−13.71	−14.70
2,3-diMe-2-Butene	1699.47	47.27	1746.74	−16.78	−16.74	−16.68
(Z)2-Heptene	1982.72	52.57	2035.29	−16.49	−16.6	−16.9
(E)2-Heptene	1982.72	53.53	2036.25	−17.44	−17.6	−17.6
(Z)3-Heptene	1982.72	52.86	2035.58	−16.77	−16.4	−16.90
(E)3-Heptene	1982.72	53.69	2036.40	−17.60	−17.4	−17.60
(Z)3-Me-3-Hexene	1982.72	55.95	2038.66	−19.86	−19.83	−18.60
(E)3-Me-3-Hexene	1982.72	56.02	2038.74	−19.93	−19.83	−19.22
2,4-diMe-1-Pentene	1982.72	56.48	2039.19	−20.74	−19.50	−20.27
4,4-diMe-1-Pentene	1982.72	54.75	2037.47	−19.35	−18.17	−19.20
(E)4,4-diMe-2-Pentene	1982.72	56.98	2039.70	−21.58	−21.47	−21.46
(E)2,2-diMe-3-Hexene	2265.97	67.55	2333.52	−26.56	−26.23	−26.16
3-Et 2-Me-1-Pentene	2265.97	66.16	2332.12	−24.82	−24.55	−24.40

The assignment of the Z and E isomers of 3-Me-3-Hexene and of 3-Me-2-Pentene are those given by de Haan and van de Ven[3]. The experimental ΔH_f° values are taken from Ref. 4.

sented in this Section are basically simple (Eqs. 5.46–5.48). However, this point seems now to be overshadowed by the tedious expansion of the $\Sigma_i\Sigma_j a_{ij}\Delta q_i$ term, thus leaving us with the impression that the beauty of the theoretical simplicity has grown dim somewhere along the road. It appears in order, therefore, to sit back and to take on the problem from a fresh point of view.

4. BOND-BY-BOND CALCULATION OF OLEFINS

At this stage, it is instructive to examine the individual bonds one by one in terms of the original equation $\varepsilon_{ij} = \varepsilon^\circ_{ij} + a_{ij}\Delta q_i + a_{ji}\Delta q_j$. This makes it easier to follow what has happened during the avalanche of calculations which led to the final expression for $\Sigma_i\Sigma_j a_{ij}\Delta q_i$.

First of all, it is clear that the CH and CC bonds involving sp^3 carbons are like those of the saturated hydrocarbons, i.e.,

$$\varepsilon_{CH}(sp^3-H) = 106.806 + a_{CH}\Delta q_C + a_{HC}\Delta q_H \tag{8.14}$$

$$\varepsilon_{CC}(sp^3-sp^3) = 69.633 + a_{CC}\Delta q_{Ci} + a_{CC}\Delta q_{Cj} \tag{8.15}$$

where a_{CC}, a_{CH} and a_{HC} are the appropriate "alkane parameters" given in Table 5.7. Of course, Δq_C and Δq_H are expressed with respect to the carbon and hydrogen net charges of ethane, which is the molecule of reference used in the definition of ε°_{CH} and ε°_{CC} (indicated above in kcal mol^{-1}).

The carbon-carbon double bond energy, on the other hand, is

$$\varepsilon_{CC}(sp^2-sp^2) = 139.27 + a_{C=C}\Delta q_{Ci} + a_{C=C}\Delta q_{Cj} \tag{8.16}$$

In this case, the Δq_C's are expressed relative to the ethylene carbon net charge and the $\varepsilon^\circ_{C=C}$ value (given here in kcal mol^{-1}) is for ethylene. Taking into account that the charges at vinyl carbon atoms vary at both the σ and π levels (i.e., $\Delta q_C = \Delta q_\sigma + \Delta q_\pi$ with $\Delta q_\sigma = m\Delta q_\pi$), we deduce the appropriate average $a_{C=C}$ parameter from Eq. 5.50. The fact that $a_{C=C}$ ($\simeq -0.291$ a.u.) turns out to be negative indicates that any gain in total electronic charge at the vinyl carbon atoms has a stabilizing effect as regards the double bond they form.

We can now proceed with the evaluation of the bonds vinyl carbons form with hydrogen and sp^3 carbon atoms, keeping in mind that charge increments are always to be expressed with respect to the charges in the reference bonds. The use of $\varepsilon^\circ_{CH} = 106.806$ and $\varepsilon^\circ_{CC} = 69.633$ kcal mol^{-1} (defined for ethane) creates a difficulty because it is more practical to select ethylene, rather than ethane, as reference molecule for defining charge increments at vinyl carbon atoms. For this reason, it appears convenient to evaluate once and for all the appropriate change of reference, i.e., the effects due to the gain in electron population ($\Delta q^\circ_{C\alpha} = -27.4$ me) accompanying the change $C(sp^3) \to C(sp^2)$ from ethane to ethylene. For CH bonds, this change represents a gain in stability of $\langle a_{C_\alpha H}\rangle \Delta q_{C_\alpha}^\circ = 3.872$ kcal mol^{-1} (Eq. 8.7). The corresponding gain in CC bond energy, $\langle a_{C_\alpha C}\rangle \Delta q_{C_\alpha}^\circ = 8.053$ kcal mol^{-1}, is estimated in the same fashion, i.e.,

$$\langle a_{C_\alpha C}\rangle \Delta q_{C_\alpha}^\circ = \left[\frac{1}{2}(a_{CC} + a_{C_\alpha C}^\sigma)\right]\Delta q_{C_\alpha}^\circ - \frac{3}{7}0.529\left(\frac{4}{1.53} - \frac{4}{1.55}\right) \text{ a.u.}$$

with $a_{CC} = -0.777$ and $a_{C_\alpha C}^\sigma = -0.718$ a.u. In this manner we deduce that

$$\varepsilon_{CH}(sp^2\text{–}H) = 110.68 + a_{C_\alpha H}\Delta q_{C_\alpha} + a_{HC}\Delta q_H, \tag{8.17}$$

$$\varepsilon_{CC}(sp^2\text{–}sp^3) = 77.69 + a_{C_\alpha C}\Delta q_{C_\alpha} + a_{CC}\Delta q_C. \tag{8.18}$$

The clear advantage offered by Eqs. 8.17 and 8.18 rests in the fact that the charge increment at the vinyl carbon atom, Δq_{C_α}, is now most conveniently expressed relative to ethylene. Of course, Δq_H and Δq_C (for sp^3 carbon) are still taken with respect to the corresponding ethane net charges. The appropriate $a_{C_\alpha H} = 0.723$ and $a_{C_\alpha C} = 0.438$ a.u. values are derived from Eq. 5.50, as indicated in Section 2.1.

With this change of reference, we are back to square one in that energy calculations can, again, be kept at their most elementary level using the bond-by-bond approach. More importantly, however, the simplicity of the method is not obscured by charge normalization constraints and the appropriate change of reference included in Eqs. 8.17 and 8.18, all of which play a role in the derivation of the sum $\Sigma_i\Sigma_j a_{ij}\Delta q_i$. An instructive example is that of ethylene (energies in kcal mol^{-1}): Eq. 8.16 gives $\varepsilon_{C=C} = 137.27$ (because the Δq_C's are 0) and Eq. 8.17 gives (with $q_H = -3.85$ and, thus, $\Delta q_H = 7.85$ me, as deduced in Section 2.2) $\varepsilon_{CH} = 105.72$. It follows that $\Delta E_a^* = 562.15$ (experimental: 562.22, from Eq. 5.5). This example illustrates the validity of Eq. 8.7 for estimating $\langle a_{C_\alpha H}\rangle \Delta q_{C_\alpha}^\circ$ (Section 2.3). The bond-by-bond calculation of tetramethylethylene is equally instructive. Using the charges deduced in Section 2.2, we obtain $\varepsilon_{CH} = 107.703$, $\varepsilon_{C=C} = 139.25$, $\varepsilon_{CC}(sp^2\text{–}sp^3) = 78.74$ and, hence $\Delta E_a^* = 1746.65$ (experimental: 1746.64, from Eqs. 5.5 and 7.6). This result illustrates the validity of the approximation used in the estimate of $\varepsilon_{CC}^\circ(sp^2\text{–}sp^3) = 69.633 + \langle a_{C_\alpha C}\rangle \Delta q_{C_\alpha}^\circ = 77.69$ kcal mol^{-1} which is part of Eq. 8.18.

While these calculations serve as examples for the simple bond-by-bond approach, they are particularly useful in another respect. It is clear that the location of the center of charge of $2p$ electrons engaged in a π bond, i.e., their shift toward the other carbon participating in the double bond, plays a crucial role in the calculation of $\langle a_{C_\alpha H}\rangle \Delta q_{C_\alpha}^\circ$ and $\langle a_{C_\alpha C}\rangle \Delta q_{C_\alpha}^\circ$ and that the results deduced for ethylene and tetramethylethylene depend decisively on the shift which is assumed. Here we may consider that overlapping p electrons forming a π bond are expected, in keeping with current views, to displace their respective centers of charge inwards, as bonding orbitals tend to collect the charge between the bonded atoms, whereas populated corresponding antibonding orbitals would have the opposite effect. It is, therefore, fair to say that the proposed model is certainly in qualitative accord with the accepted understanding of π bonds. It remains that the shift of ~ 0.04 Å which is assumed here is a product of trial and error analysis. Its appropriateness can be illustrated as follows. If no shift is considered (which would clearly represent an arbitrary assumption), we obtain $\langle a_{C_\alpha H}\rangle \Delta q^\circ_{C_\alpha} = 6.267$ and $\langle a_{C_\alpha C}\rangle \Delta q^\circ_{C_\alpha} = 12.852$ kcal mol^{-1}. As a consequence, the ΔE^*_a's calculated for ethylene and tetramethylethylene would be in error by 9.58 and 19.20 kcal mol^{-1}, respectively. Clearly, a shift of ~ 0.04 Å is the only value which satisfies simultaneously ethylene and tetramethylethylene.

In this matter, it is also instructive to look at the general formula giving $\Sigma_i\Sigma_j a_{ij}\Delta q_i$ (Eq. 8.12, viz. 8.13). If no shift is considered, the coefficient $\lambda^\sigma_1\Delta q^\circ_{C_\alpha} + \lambda^\sigma_1 - \lambda^\pi_1$ of $\Sigma^\alpha N_{C_\alpha C}$ and the constant term (Eq. 8.8) would amount to 6.59 and -9.56 kcal mol^{-1}, respectively. With a shift of 0.06, instead of 0.04 Å, these two parameters would become 2.87 and -23.80 kcal mol^{-1}, respectively. These examples and inspection of Eq. 8.12 (viz. 8.13) indicate that any choice other than ~ 0.04 Å leads to two distinct and important sources of systematic errors. While, of course, we would prefer to obtain this result from strict theoretical considerations, it is obvious from the present calculations that there is not much room left for error in locating the "effective" center of charge of p electrons engaged in a vinyl π bond.

5. EXO- VS. ENDOCYCLIC DOUBLE BONDS

The bond-by-bond calculations offered above have the merit of showing that, basically, olefins can be treated as simply as saturated hydrocarbons. The general solution for $\Sigma_i\Sigma_j a_{ij}\Delta q_i$, however, has merits of its own, which are exploited as follows.

To begin with, *the term in* $\Sigma N_{C_\alpha C}$ *indicates an* a priori *gain in stability of* ~ 4 *kcal mol^{-1} for each* CC *bond formed by a vinyl carbon, in replacement for a* CH *bond.* This aspect is, of course, included in the calculations presented in Table 8.1. It appears, however, that *this gain in stability is somewhat reduced by electron redistributions.* For example, in going from 2 methyl-

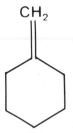

Methylenecyclohexane
ΔH°_f (298.15, gas) = -7.2
kcal mol^{-1}. Calcd. (Benson[4]):
-8.1 kcal mol^{-1}.

1-Methylcyclohexene
ΔH°_f (298.15, gas) = -10.0
kcal mol^{-1}. Calcd. (Benson[4]):
-9.7 kcal mol^{-1}.

1-pentene ($\Sigma N_{C\alpha C} = 2$) to 3 methyl-1-pentene ($\Sigma N_{C\alpha C} = 1$), the decrease in ΔE_a^* is not ~ 4 but only 2.6 kcal mol^{-1} because of compensating electronic rearrangements, but the leading term is clearly given by $\sim 4\Sigma N_{C\alpha C}$. On these grounds it is easy to predict that methylenecyclohexane ($\Sigma N_{C\alpha C} = 2$) is less stable than 1-methylcyclohexene ($\Sigma N_{C\alpha C} = 3$).

This rationale is relevant in the discussion of the relative thermochemical stabilities of exo- *vs.* endocyclic double bonds. Of course, the fact alone that a double bond is exocyclic does not necessarily imply a loss in stability. For example, ethylidenecyclohexane and 1-ethylcyclohexene, for which $\Sigma N_{C\alpha C} = 3$ in both cases, are similar as regards their thermochemical stabilities.

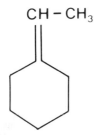

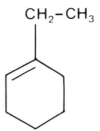

Ethylidenecyclohexane
ΔH_f° (298.15, gas)[6] $= -14.7$
kcal mol^{-1}.

1-Ethylcyclohexene
ΔH_f° (298.15, gas)[4] $= -15.0$
kcal mol^{-1}. Calcd. (Benson[4]):
-14.5 kcal mol^{-1}.

These theoretical expectations, based on the $\sim 4 \Sigma N_{C\alpha C}$ term, fully confirm similar views expressed by Fuchs and Peacock[6], suggesting that the cyclic compounds differ in double bond substitution rather than conformational stability. This is an area certainly worth additional investigation.

The same trends are also observed for the five-membered ring analogs, as revealed by their heats of formation (gas phase, 298.15 K, indicated below in kcal mol^{-1}). Indeed, 1-methylcyclopentene (-0.86[6]) with $\Sigma N_{C\alpha C} = 3$ is significantly more stable than methylenecyclopentane (2.4[4]) with $\Sigma N_{C\alpha C} = 2$. In turn, 1-ethylcyclopentene (-4.72[6]) and ethylidenecyclopentane (-4.33[6]), with $\Sigma N_{C\alpha C} = 3$ in both cases, are similar in thermochemical stability. Of course, a more detailed analysis would require the precise knowledge of the charge effects in five-membered ring compounds—a topic which still awaits investigation. While this sort of similarity between 5- and 6-membered cyclic hydrocarbons is not surprising, it must be added that the situation is quite different in the 3-membered ring analogs, methylenecyclopropane (48.0[4]) being clearly more stable than 1-methylcyclopropene (58.2[4]). It seems obvious that the considerable additional "strain" in the latter isomer, due to the short endocyclic double bond, should be pinpointed as the factor responsible for this reversal in the order of stabilities, overriding any anticipated "$\Sigma N_{C\alpha C}$ effect".

6. DIENES, ALLENES, ALKYNES AND BENZENE

Detailed charge analyses are not yet available for these classes of compounds, thus precluding temporarily the evaluation of the $\Sigma_i\Sigma_j a_{ij}\Delta q_i$ terms which are required in any accurate calculation of their atomization energies. However, it is still possible to gain some insight into the major factors governing their thermochemical stabilities, in the following manner. So far we have focussed on the $\Sigma_i\Sigma_j a_{ij}\Delta q_i$ part of the expression $\Delta E_a^* = \Sigma\varepsilon_{ij}^\circ + \Sigma_i\Sigma_j a_{ij}\Delta q_i + \Delta E_{nb}^*$. This is, indeed, the energy term describing isodesmic structural changes at the vibrationless level if we agree upon neglecting here minor changes in nonbonded interactions. It is now interesting to examine the $\Sigma\varepsilon_{ij}^\circ$ part, remembering that the individual ε_{ij}° bond contributions are those defined in Eq. 5.35, certainly not to be mistaken for empirical "apparent" bond energies like the $\varepsilon_{CC}^{appar.} \simeq 80.7$ and $\varepsilon_{CH}^{appar.} \simeq 105.0$ kcal mol^{-1} figures discussed in Chapter 6.5 or with an empirical CC double bond energy $\varepsilon_{C=C}^{appar.} = 139.27 - 19.06 \simeq 120.2$ kcal mol^{-1} which would be appropriate for mono-olefins (see Eq. 8.12). Note that these "apparent" bond energies correspond to the conventional empirical ones which attempt to approach experimental ΔE_a^*'s as closely as possible by means of appropriate sets of constant bond contributions; in the case of ethylenic double bonds, their conventional empirical evaluation automatically counts the term given by Eq. 8.8 together with $\varepsilon_{C=C}^\circ$, because this term occurs once for each double bond like those considered here. This sort of error is avoided in the present theory with the use of the appropriate ε_{ij}°'s given by Eq. 5.35.

First of all, an important conclusion derived from the detailed energy analyses of alkanes and simple olefins is that *the energy contribution of a carbon-carbon double bond is nearly twice that of a simple bond* although, admittedly, minor uncertainties subsist as regards the precision of the $\varepsilon_{C=C}^\circ$ term calculated in this study. In the following, we shall tentatively assume that, indeed, $\varepsilon_{C=C}^\circ = 2\varepsilon_{CC}^\circ$ and see what happens. The pertinent observation is that the $\Sigma_i\Sigma_j a_{ij}\Sigma q_i$ terms are quite similar for the alkanes and the alkenes having the same number of carbon atoms, except for ethane $vs.$ ethylene. This information is easily extracted from the results given in Tables 6.1 and 8.1. If we decide not to be too critical about discrepancies of a few kcal mol^{-1} when examining the major trends in hydrocarbon chemistry, it appears that the difference in ΔE_a^* energy between a C_nH_{2n+2} and a C_nH_{2n} unsaturated hydrocarbon is roughly due to the removal of two CH bonds (2×106.81 kcal mol^{-1}) and their replacement by a CC bond (69.63 kcal mol^{-1}), representing a decrease of ~ 144.0 kcal mol^{-1} in ΔE_a^* energy. Table 8.2 offers this type of rough estimates using, where appropriate, the averages of the various alkene isomers (e.g., the average ΔE_a^* of cis- and trans-2-butene, in the comparison with butane).

These rough estimates, whose theoretical validity rests on the similarity in the $\Sigma_i\Sigma_j a_{ij}\Delta q_i$ values for alkanes and alkenes with the same number of carbon atoms, certainly support the result that $\varepsilon_{C=C}^\circ \simeq 2\varepsilon_{CC}^\circ$. Of course, this suggests

TABLE 8.2. Alkene Atomization Energies Estimated from the Parent Alkanes (kcal mol^{-1})

	ΔE_a^*, Alkene	
Alkane/Alkene	Estd.	Exptl.
Propane/propene	860.1	858.56
Butane/butene	1154.2	1154.5
Pentane/pentene	1448.2	1447.6
Hexane/hexene	1742.0	1741.9
Cyclohexane/cyclohexene	1616.8	1616.1[a]
Methylcyclohexane/1-methylcyclohexene	1913.1	1913.2[b]
Cyclohexane/1,4-cyclohexadiene	1472.8	1472.8[c]
Butane/1,3-butadiene	1010.2	1010.15[d]

[a] Calculated (Eq. 5.5) from ΔH_f°(298.15, gas)[4] $= -0.84$ and ZPE $+ H_T$-$H_0 = 92.70$ kcal mol^{-1} (from data given in Ref. 8). [b] Deduced from ΔH_f°(298.15, gas)[4] $= -10.0$ and ZPE $+ H_T$-$H_0 = 92.70 + 18.21 - 0.34 = 110.57$ kcal mol^{-1}, following the rules indicated in Chapter 7. [c] From ΔH_f°(298.15, gas)[4] $= 26.3$ and ZPE $+ H_T$-$H_0 = 77.8$ kcal mol^{-1} [estimated from the rules indicated in Chapter 7 using the values of butene and ethylene less $23.7 + 4RT$ (giving 77.75 kcal mol^{-1}) and from the difference between cyclohexane and cyclohexene (14.84 kcal mol^{-1}) which is subtracted from the value of cyclohexene (giving 77.86 kcal mol^{-1})]. [d] Deduced from ΔH_f°(298.15, gas)[4] $= 26.33$, H_T-$H_0 = 3.63$ (Ref. 9) and ZPE $= 51.56$ kcal mol^{-1} (deduced from data given in Ref. 10).

TABLE 8.3. Hartree–Fock Calculations of Selected Bond Energy Contributions (Eqs. 5.33 and 5.35) Using the Contracted $[5s3p|3s]$ Basis Set

Molecule	V(C, mol), a.u.	ε_{CC}, kcal mol^{-1}	Ratio Relative to 66.7
Ethane	-88.4670	66.7	1.00
Ethylene	-88.4168	133.7	2.00
Acetylene	-88.3345	193.6	2.90
Benzene	-88.3391	108.1	1.62
Allene	-88.3582 (C–1)	128.5	1.93
	-88.2815 (C–2)		

The corresponding HF values were used for atomic carbon, i.e., E(C, at) $= -37.5995$ a.u. and V_{ne}(C, at) $= -87.9315$ a.u., as well as $\partial\varepsilon_{CH}/\partial Z_C \simeq 0.026$ a.u.

the question about triple bonds and the CC bonding in benzene. In order to gain an insight into these problems, we go back to the fundamental equations 5.33 and 5.35 and calculate the appropriate bond energies from the total potential energies $V(k, \text{mol})$ at the atomic nuclei. The following results (Table 8.3) are derived from Hartree–Fock calculations[7] at the $(9s5p|6s) \rightarrow [5s3p|3s]$ level. While this basis cannot match experimental accuracy, it remains that comparisons on a relative scale can be regarded as meaningful. While the result $\varepsilon_{C=C}^\circ \simeq 2\varepsilon_{CC}^\circ$ is clearly confirmed, it also appears that the carbon-carbon triple bond in acetylene is not far from being the triple of a single bond contribution. It remains, however, that the 2.9 factor is significant. Indeed, by subtracting 4 CH bond contributions from butane and adding 1.9 ε_{CC}° (i.e., by subtracting ~ 295 kcal mol^{-1} from the ΔE_a^* of butane) we obtain, in this rough estimate, ΔE_a^* (2-butyne) $\simeq 1003.1$ kcal mol^{-1}. Under these circumstances, the agreement with the experimental value[11], $\Delta E_a^* = 1001.5$ kcal mol^{-1}, is certainly satisfactory. Similarly, using now the 1.6 factor deduced for benzene, we estimate its atomization energy from that of cyclohexane by removing 6 CH bonds and adding 6 times $0.6\varepsilon_{CC}^{\circ}$. This estimate leads to ΔE_a^* (benzene) $\simeq 1370.6$ kcal mol^{-1}, still in rough agreement with its experimental value[13], $\Delta E_a^* = 1366.51$ kcal mol^{-1}. Finally, taking propane as a model compound in order to evaluate allene, we replace 4 CH bonds by 2 CC bonds using the 128.5/66.7 ratio given by HF calculations. The value estimated for allene, $\Delta E_a^* \simeq 705.9$ kcal mol^{-1}, is reasonably close to its thermochemical counterpart deduced[14] from Eq. 5.5, ~ 701.9 kcal mol^{-1}. Similarly, the atomization energy of 2,3-pentadiene, 1294.0 kcal mol^{-1}, deduced from n-pentane compares favorably with the result derived from Eq. 5.5[15], 1293.3 kcal mol^{-1}. These rough estimates illustrate, in a simple manner, the general validity of bond energies deduced from Eq. 5.35 using Hartree–Fock potential energies, particularly as regards their relative ordering.

No claim for accuracy can be made for this sort of brute force skeletal approach. It remains, however, that a link is created in this manner between typical saturated, olefinic, acetylenic and aromatic compounds. Of course, a host of details have to be worked out, a task which looks promising. The present results should certainly encourage us to do some more thinking about possible further applications of Eq. 5.35.

7. CONCLUSIONS

In the first place, it is important to assess the significance of the final numerical results used in comparisons between theory and experiment. Atomization energies of olefins calculated from ^{13}C NMR spectra for hypothetical vibrationless molecules, on one hand, and ZPE + H_T-H_0 energies "constructed" from appropriate rules, on the other, enable a reasonable evaluation of standard enthalpies of formation of the olefins in spite of the fact that, in general, spectroscopic information is scarce for this class of compounds. The quality of the comparisons offered for the olefins (Table 8.1) should be judged accordingly, also keeping in mind that non-bonded energies were not calculated explicitly. Results derived from the group additivity scheme, as originally proposed by Benson and Buss[16], are also indicated. The good quality of this additivity scheme certainly justifies Benson's statement that sufficient data are available for hydrocarbons to render suspect any data which deviate from the value predicted by the scheme by more than ~ 1 kcal mol^{-1} in ΔH_f°. It is noteworthy that in the four cases showing the largest discrepancies (0.71–1.26 kcal mol^{-1}) between our theoretical ΔH_f°'s and their experimental counterparts, Benson's values are, indeed, very similar to the present theoretical results. Under these circumstances it appears fair to conclude that the overall agreement (0.30 kcal mol^{-1} average deviation) between theory and experiment is similar in quality to that obtained for the saturated hydrocarbons.

Bond-by-bond calculations using the basic equation $\varepsilon_{ij} = \varepsilon_{ij}^\circ + a_{ij}\Delta q_i + a_{ji}\Delta q_j$ are no more difficult in the case of ethylenic hydrocarbons than for saturated hydrocarbons. The difference in behavior between σ and π electrons in the bonds involving sp^2 carbon atoms is taken care of in the a_{ij}'s: these are calculated separately for σ and π electrons in the usual way, from Eq. 5.47, and the appropriate averages $a_{ij} = (ma_{ij}^\sigma + a_{ij}^\pi)/(m + 1)$ (Eq. 5.50) are taken, thus accounting for the inverse variations, $\Delta q_\sigma = m\Delta q_\pi$, of σ and π charges. The fact that $a_{C=C}$, refering to the sp^2–sp^2 CC bond, is negative (-0.291 a.u.) indicates that any increase in total ($\sigma + \pi$) electron population at an sp^2 carbon promotes a stabilization as concerns the contribution of the double bond itself. Contrasting with all the a_{ij}'s encountered so far, however, those of an sp^2 carbon forming a CH bond ($a_{C\alpha H} = 0.723$ a.u.) or a CC bond with an sp^3 carbon ($a_{C\alpha C} = 0.438$ a.u.) are positive, meaning that a gain in total ($\sigma + \pi$) charge at the sp^2 carbon actually leads to a lowering in stability. This behavior is readily understood. Indeed, in this case a gain in total electronic charge is due to an increase in π population prevailing over an almost equally important decrease in σ population: the loss of σ electrons dictates the overall behavior of the CH and CC bonds concerned, whose contributions to the molecular stability are lowered.

The general formula describing C_nH_{2n} olefins is obtained from the sum of all the individual bond contributions. Its derivation, which has been presented in a detailed step-by-step approach, includes charge normalization in order to eliminate the terms in hydrogen charges, thus yielding an expression in terms of carbon charges only. The latter are conveniently expressed relative to ethane and ethylene for the sp^3 and sp^2 carbon atoms, respectively, so that the final result can be given in terms of the corresponding ^{13}C NMR shifts. Of course, this is the most practical way of obtaining the ΔE_a^* atomization energies because all the intricacies related to charge analyses are already taken care of. Moreover, inspection of the general formula facilitates general conclusions to be drawn. An important contribution is due to the alkyl groups: the corresponding $\lambda_1\Sigma N_{CC}\delta_C + \lambda_2\Sigma\delta_C$ term is identical to that encountered in the study of saturated hydrocarbons (Chapter 6). A noteworthy point concerns the $\lambda_3\Sigma\delta_{C\alpha}$ term describing the charge effects of sp^2 carbon atoms. The fact that λ_3 is positive (as is the case with λ_1 and λ_2) indicates that *downfield ^{13}C shifts of sp^2 carbons reflect a gain in molecular stability*. Contrasting with sp^3 carbons, however, a downfield shift observed for an sp^2 carbon indicates a decrease of its total electron population. Now we know that such a decrease represents a lowering in π population prevailing over an almost equally important increase in σ population. *The gain in stability accompanying downfield shifts of sp^2 carbons is thus traced back to a prevailing role of σ charges*, whose population increases, *in promoting the stability of an alkene molecule.* Briefly, a gain of electronic charge in the double bond system (revealed by upfield shifts) is an unfavorable situation from the point of view of molecular stability because it promotes the formation of π electrons at the expense of σ electrons. This observation holds although the double bond itself has gained in stability. Finally, the term in ΣN_{CC} indicates that the thermochemical stability of an alkene increases with an increasing number of CC, rather than CH, bonds formed by the two sp^2 carbons—an observation which is important in discussions concerning the relative stabilities of exo- vs. endocyclic double bonds.

Detailed energy analyses of unsaturated hydrocarbons other than the $R_1R_2C{=}CR_3R_4$ olefins are presently unavailable because the knowledge concerning their charges is insufficient. However, rough estimates can be offered by means of the basic bond energy formula (Eq. 5.35) and Hartree–Fock calculations of potential energies at the individual nuclei. It appears that the carbon–carbon bond energy contributions of ethane, ethylene, acetylene, and benzene are in a ratio of $1 : 2 : 2.9 : 1.6$—information which permits a rapid estimate of approximate atomization energies of unsaturated hydrocarbons from those of their corresponding saturated parent compounds and, more importantly, shows the links relating the various classes of saturated and unsaturated hydrocarbons in a unifying fashion.

REFERENCES

1. M.-T. Béraldin and S. Fliszár, *Can. J. Chem.*, **61**, 197 (1983); **61**, 1291 (1983).
2. J.B. Stothers, "Carbon–13 NMR Spectroscopy", Academic Press, New York, NY, 1972 (C_2H_4 to C_4H_8); G.J. Abruscato, P.D. Ellis, and T.T. Tidwell, *J. Chem. Soc., Chem. Comm.*, 988 (1972) (trimethylethylene); J.W. de Haan, L.J.M. van de Ven, A.R.N. Wilson, A.E. van der Hout–Lodder, C. Altona, and D.H. Faber, *Org. Magn. Res.*, **8**, 477 (1976) (1-olefins).
3. J.W. de Haan and L.J.M. van de Ven, *Org. Magn. Res.*, **5**, 147 (1973).
4. S.W. Benson, F.R. Cruickshank, D.M. Golden, G.R. Haugen, H.E. O'Neal, A.S. Rodgers, R. Shaw, and R. Walsh, *Chem. Rev.*, **69**, 279 (1969).
5. H. Henry and S. Fliszár, *J. Am. Chem. Soc.*, **100**, 3312 (1978).
6. R. Fuchs and L.A. Peacock, *Can. J. Chem.*, **57**, 2302 (1979).
7. S. Fliszár, M. Foucrault, M.-T. Béraldin, and J. Bridet, *Can. J. Chem.*, **59**, 1074 (1981); M. Foucrault, M.Sc. Thesis, Department of Chemistry, Université de Montréal, Montréal (1981).
8. N. Neto, C. diLauro, E. Castellucci, and S. Califano, *Spectrochim. Acta*, **23A**, 1763 (1967).
9. F.D. Rossini, K.S. Pitzer, R.L. Arnett, R.M. Braun, and G.C. Pimentel, "Selected Values of Physical and Thermodynamic Properties of Hydrocarbons and Related Compounds", American Petroleum Institute, Carnegie Press, Pittsburgh, PA, 1953.

10. Yu. N. Panchenko, *Spectrochim. Acta*, **31A**, 1201 (1975).
11. $\Delta H_f^{\circ}(298.15, \text{gas}) = 34.97$ kcal mol^{-1} (from Ref. 4); $H_T\text{-}H_0 = 3.96$ kcal mol^{-1} (from Ref. 9); ZPE = 51.21 kcal mol^{-1}, from the data given in Ref. 12.
12. G. Herzberg, "Molecular Spectra and Molecular Structure. II. Infrared and Raman Spectra of Polyatomic Molecules", D. Van Nostrand Co. Inc., Princeton, New Jersey, 1968.
13. $\Delta H_f^{\circ}(298.15, \text{gas}) = 19.82$ kcal mol^{-1} (from Ref. 4); ZPE $+ H_T\text{-}H_0 = 66.22$ kcal mol^{-1}, from data given in Ref. 12.
14. $\Delta H_f^{\circ}(298.15, \text{gas}) = 45.92$ kcal mol^{-1} (from Ref. 4). The ZPE $+ H_T\text{-}H_0 \simeq 37.2$ kcal mol^{-1} value is estimated from the difference between propane and propene (14.38 kcal mol^{-1}), which is subtracted from the value of propene. Compare with the Footnote at the bottom of Table 8.2.
15. $\Delta H_f^{\circ}(298.15, \text{gas}) = 31.79$ kcal mol^{-1} (from Ref. 4). The value for ZPE $+ H_T\text{-}H_0 = 73.15$ kcal mol^{-1} is estimated from the difference between *n*-pentane and pentene (14.84 kcal mol^{-1}), which is subtracted from the value of pentene. Compare with the Footnote at the bottom of Table 8.2.
16. S.W. Benson and J.H. Buss, *J. Chem. Phys.*, **29**, 546 (1958).

CHAPTER 9

Energy Analysis of Oxygen-Containing Compounds

1. INTRODUCTION

Numerical verifications of the bond energy theory have been carried out at the most detailed level for saturated hydrocarbons. The only parameter which was not derived directly from theory is the absolute magnitude of the carbon net charge in ethane, $q_C^\circ = 35.1$ me. This result is carried over in any calculation involving alkyl groups, e.g., in the study of ethylenic compounds and, in the present case, of selected ethers and carbonyl compounds. An important aspect has emerged from systematic applications of the results obtained for saturated hydrocarbons, namely, the possibility of constructing $ZPE + H_T\text{-}H_0$ energies in a simple manner. The fact that over 100 hydrocarbons, both saturated and unsaturated, were successfully calculated certainly supports the idea that vibrational energies can be constructed from appropriate additivity rules. There is perhaps an element of surprise in witnessing the accuracy which can be achieved in this way. While this type of result is, in itself, a possible source of future developments in the field of molecular vibrational energies, it appears now safe to take advantage of the confidence which has been built up in "constructed" $ZPE + H_T\text{-}H_0$ energies. Indeed, available spectroscopic data for ethers and carbonyl compounds are scarce: in fact, they are barely sufficient for the derivation of reasonably reliable $ZPE + H_T\text{-}H_0$ energies for these classes of compounds. However, the uncertainties introduced in this fashion do not obscure the significance of the comparisons between theoretical and experimental values, because the differences in ΔE_a^* energies in series of isomers largely exceed the minor uncertainties associated with the evaluation of vibrational energies by additivity rules.

Fortunate circumstances facilitate the calculation of theoretical ΔE_a^* energies of ethers. Mulliken atomic populations of the carbon atoms adjacent to oxygen vary at the $2s$ level, much in the same way as in saturated hydrocarbons, with $2p$ populations increasing by ~ 30–32 me for each hydrogen atom added to these carbons (see p. 60). On these grounds it appears reasonable to treat the α-carbons in ethers like alkane carbons as regards the partitioning of CH overlap populations, and to apply the relationship $\Delta q_C = -0.148\delta_C$ which is valid for typical alkane carbon atoms. Of course, it is necessary to define Δq_C and δ_C in an appropriate manner, namely, by taking an ether α-carbon as reference. For the oxygen charges, on the other hand, extensive *ab initio* calculations[1] (Chapter 4.5) indicate variations in electron populations occuring only at the $2p$ level and a shift $vs.$ charge slope which is $\sim 1/1.8$ that observed for sp^3 carbons, giving the relationship $\Delta q_O = -0.267\delta_O$ (Eq. 4.11), in me units. In this manner, all charge results, including those for oxygen, are anchored relative to the carbon net charge in ethane, 35.1 me. Using these charge–shift relationships, as well as charge normalization for calculating hydrogen net charges, straightforward applications of the basic energy expression $\varepsilon_{ij} = \varepsilon_{ij}^\circ + a_{ij}\Delta q_i + a_{ji}\Delta q_j$ are easily performed, as shown in a detailed example worked out in the next Section. The derivation of a general formula for $\Sigma_i\Sigma_j a_{ij}\Delta q_i$ describing dialkylethers is also presented. Considering that no empirical parameter has been introduced in carrying out these calculations for the ethers, the comparison between calculated and experimental atomization energies may be regarded as a fair test for the theory although, evidently, minor approximations still remain open for discussion.

Unfortunately, this is not the case with carbonyl compounds. The estimates of their ZPE $+ H_T$-H_0 energies from additivity rules (Chapter 7.6) are probably as reasonable as those made for the other classes of compounds which have been investigated up to this point. The real problem rests with the evaluation of the carbonyl carbon and oxygen net charges. Standard STO–3G population analyses indicate downfield NMR shifts with increasing electron populations at the carbon atoms (like in the alkane series), as indicated in Table 4.7 and Figure 4.4, but the precision of these calculations does not permit a clear assessment regarding the proper way of treating carbonyl carbon atoms. In principle, we should probably apply Eq. 5.49 by considering the individual variations of $2s$ and $2p$ populations. In practice, we use tentatively the usual charge–shift correlation, $\Delta q_C = -0.148\delta_C$ (me), in spite of the lack of adequate theoretical support. For the oxygen atoms, it appears that our information is even poorer, except as regards the general trend indicating upfield ^{17}O NMR shifts with increasing electron populations. The main difficulty in the theoretical charge analyses comes from the fact that we simply do not know how to deal with the problem of overlap populations in C=O systems. Under these circumstances, we have used the approximation $\Delta q_O \simeq 2.7\delta_O$ for carbonyl oxygens, which, of course, is entirely empirical and may in part compensate for errors made in the evaluation of a_{OC}. It is clear, therefore, that the numerical agreement between calculated and experimental energies of carbonyl compounds cannot be considered on the same footing as the verifications involving hydrocarbons and ethers. It may also be added, however, that the largest charge-dependent contributions are due to the $\lambda_1 \Sigma N_{CC} \Delta q_C + \lambda_2 \Sigma \Delta q_C$ terms describing the alkyl parts, which are now well-known. The uncertainties discussed here affect only a portion of $\Sigma_i \Sigma_j a_{ij} \Delta q_i$ which, although important at the level of experimental accuracy, does not represent the leading term of the charge-dependent part of ΔE_a^*. For this reason, the comparisons offered for carbonyl compounds still present positive aspects, moreover as the quality of the results supports the practical validity of the final energy formula.

2. CHARGE NORMALIZATION

The following considerations apply to both ethers and carbonyl compounds. For the hydrogen atoms, we deduce the sum $\Sigma \Delta q_H$ from the charge normalization constraint $\Sigma_i q_i = 0$, i.e.,

$$q_H = -\sum q_{C\alpha} - \sum q_{C \neq \alpha} - \sum q_O$$

where C_α and $C_{\neq \alpha}$ stand for carbon atoms adjacent to, $viz.$ nonadjacent to, oxygen and $q_{C\alpha}$, $viz.$ $q_{C \neq \alpha}$, are the corresponding net atomic charges. The ether or carbonyl oxygen net charge is q_O. The number of atoms of the species i is n_i.

When evaluating charge increments Δq_i it is important to consider, first, which is the appropriate reference charge and, second, that the latter is defined for an appropriately selected reference bond. For hydrogen atoms engaged in CH bonds, the reference bond energy is that of ethane, ε_{CH}°, and the corresponding reference charge is the ethane-H net charge, $q_H^\circ = -11.7$ me. Consequently, with $\Delta q_H = q_H - q_H^\circ$,

$$\sum \Delta q_H = -\sum q_{C\alpha} - \sum q_{C \neq \alpha} - \sum q_O - n_H q_H^\circ.$$

The carbon atoms other than in α position are engaged only in CC and CH bonds, which are calculated with reference to the ethane CC and CH bond energy contributions, i.e., ε_{CC}° and ε_{CH}°, respectively. The appropriate reference net charge for carbon is thus, as usual, the ethane-C net charge, $q_C^\circ = 35.1$ me. It follows therefrom, with $\Delta q_{C \neq \alpha} = q_{C \neq \alpha} - q_C^\circ$, that

$$\sum \Delta q_H = -\sum \Delta q_{C \neq \alpha} - \sum q_{C\alpha} - \sum q_O - (n_H q_H^\circ + n_{C \neq \alpha} q_C^\circ).$$

Similarly, if q_O° is the oxygen reference charge defined by selecting acetone for the carbonyl compounds or diethylether for the ethers, we write $\Delta q_O = q_O - q_O^\circ$. Finally, defining $q_{C\alpha}^\circ$ for acetone or diethylether, whichever is the appropriate reference, we write $\Delta q_{C\alpha} = q_{C\alpha} - q_{C\alpha}^\circ$ and obtain[1]

$$\sum \Delta q_H = -\left(\sum \Delta q_{C\alpha} + \sum \Delta q_{C \neq \alpha} + \sum \Delta q_O\right) - (n_{C\alpha} q_{C\alpha}^\circ + n_O q_O^\circ)$$
$$- (n_{C \neq \alpha} q_C^\circ + n_H q_H^\circ). \tag{9.1}$$

Equation 9.1 enables the evaluation of $\Sigma \Delta q_H$ from the knowledge of the $\Delta q_{C\alpha}$'s, $\Delta q_{C \neq \alpha}$'s, and the Δq_O's. Under the circumstances indicated in Chapter 4, the latter quantities can often be deduced from the corresponding chemical shifts. At this stage it is important to note that the $\delta_{C \neq \alpha}$ shift data are to be expressed with reference to the ethane carbon atom (5.8 ppm from TMS), whereas the $\delta_{C\alpha}$ shifts are relative to the acetone carbonyl-C atom (for carbonyl compounds), viz. the α-carbon of diethylether (for the ethers), which are the atoms appropriately selected as references in the respective series of compounds. Similarly, the ^{17}O NMR shifts are relative to acetone and diethylether for the carbonyl compounds and ethers, respectively. Moreover, the sum $n_{C\alpha}q^\circ_{C\alpha} + n_O q^\circ_O$ gives $2q^\circ_{C\alpha} + q^\circ_O \simeq 67.7$ me for diethylether and $q^\circ_{C\alpha} + q^\circ_O \simeq -7.2$ me for acetone, from charge normalization (Chapter 4.6). Attention must be given to the difference

$$\Delta q^\circ_{C\alpha} = q^\circ_{C\alpha} - q^\circ_C. \tag{9.2}$$

Indeed, for CO bonds the reference is $q^\circ_{C\alpha}$ but in calculations of CC and CH bonds, including those involving α-carbons, the appropriate reference is q°_C. This is why this difference (Eq. 9.2) appears in the final expressions for $\Sigma_i \Sigma_j a_{ij} \Delta q_i$, as indicated further below. It is not possible to simply deduce $\Delta q^\circ_{C\alpha}$ from the appropriate ^{13}C NMR shifts, as oxygen appears to introduce an "extra" downfield shift at α-carbons, estimated at ~ 41.7 ppm in the case of the ethers[2]. Assuming here that this shift, which could be due in part to the electric field of the oxygen dipole, is not primarily a carbon-charge effect, we estimate the latter by subtracting tentatively 41.7 ppm from the observed α-carbon shifts (60 and 199 ppm from ethane for diethylether[2] and acetone[3], respectively). It turns out that the $\Delta q^\circ_{C\alpha} \simeq -2.7$ me (ether) and $\Delta q^\circ_{C\alpha} \simeq -23$ me (acetone) values obtained in this fashion (Eq. 4.10) are reasonable: atomization energies calculated with these $\Delta q^\circ_{C\alpha}$ values differ from their experimental counterparts by ~ 0.35 kcal/mol (average deviation). A refinement involving experimental energies yields, as described below,

$$\Delta q^\circ_{C\alpha} = -3.84 \text{ me for diethylether} \tag{9.3}$$

and

$$\Delta q^\circ_{C\alpha} = -21.1 \text{ me for acetone.} \tag{9.4}$$

Of course, not too much importance should be attached to these estimates since they may, in part, correct for minor uncertainties associated with the a_{ij}'s, with the charge–shift relationships, and with the evaluation of vibrational and thermodynamic data used in this analysis. Clearly, the difficulties are now in the evaluation of fraction of kcal/mol uncertainties, thus preventing us temporarily from further improving the calculations. This is best seen by examining experimental results.

3. ETHERS

The theoretical sums $\Sigma\varepsilon_{ij}$ can be deduced from bond-by-bond calculations of the individual ε_{ij}'s, using Eq. 5.46, in a way similar to that employed in Chapter 6.2 for the chair and boat forms of cyclohexane. This approach is illustrated by the following example. For di-n-propylether, $\delta_O = -10.0$ ppm (from diethylether) gives (Eq. 4.11) $\Delta q_O = -0.267\delta_O = 2.67$ me; $\delta_{C\alpha} = 6.8$ ppm (from the α-carbon of diethylether) gives (Eq. 4.10) $\Delta q_{C\alpha} = -0.148\delta_{C\alpha} = -1.01$ me. The Δq_O and $\Delta q_{C\alpha}$ values, relative to the reference which is diethylether, are used in the calculation of the CO contributions, giving (Eq. 5.46, Tables 5.6 and 5.7, with 1 a.u. = 627.51 kcal/mol) $\varepsilon_{CO} = 79.78 - (1.135)(627.51)(-1.01 \times 10^{-3}) - (0.804)(627.51)(2.67 \times 10^{-3}) = 79.15$ kcal/mol. For the β and γ C atoms, the shifts (relative to ethane) are 17.6 and 5.0 ppm, respectively, giving $\Delta q_{C\beta} = -0.148\delta_C = -2.60$ and $\Delta q_{C\gamma} = -0.74$ me. With these Δq_C's, Eq. 5.46 gives, for the C_β—C_γ bonds, $\varepsilon_{CC} = 69.63 - (0.777)(627.51)(-2.60 \times 10^{-3}) - (0.777)(627.51)(-0.74 \times 10^{-3}) = 71.26$ kcal/mol. For the C_α—C_β bonds, we use again $\Delta q_{C\beta} = -2.60$ me and, for the α-carbon, $\Delta q_C = q_{C\alpha} - q_C^\circ = q_{C\alpha} - q_{C\alpha}^\circ + q_{C\alpha}^\circ - q_C^\circ = \Delta q_{C\alpha} + \Delta q_{C\alpha}^\circ = -1.01 - 3.84 = -4.85$ me (from Eqs. 9.2 and 9.3). The C_α—C_β bond energy contribution is thus $\varepsilon_{CC} = 69.63 - (0.777)(627.51)(-4.85 \times 10^{-3}) - (0.777)(627.51)(-2.60 \times 10^{-3}) = 73.26$ kcal/mol. The $\Delta q_C = -4.85$ me value is also used in the calculation of the corresponding ε_{CH} contributions involving α-carbons. Moreover, we obtain from the figures given above that $\Sigma\Delta q_{C\alpha} = 2(-1.01) = -2.02$ me, $\Sigma\Delta q_{C\neq\alpha} = 2(-2.60-0.74) = -6.68$ me, and $\Sigma\Delta q_O = 2.67$ me. Finally, using $q_{C\alpha}^\circ = 31.26$, $q_O^\circ = 5.18$, $q_{C\neq\alpha}^\circ = 35.10$, and $q_H^\circ = -11.70$ me (see Chapter 4.6), it follows from Eq. 9.1 that $\Sigma\Delta q_H = -38.27$ me. The average Δq_H is therefore -2.734 me relative to ethane, for an average $a_{HC}\Delta q_H = -1.007 \times 627.51(-2.734 \times 10^{-3}) = 1.727$ kcal/mol contribution to ε_{CH}. The calculation of all the ε_{CH}'s is now straightforward and gives $106.81 - (0.394)627.51(-4.85 \times 10^{-3}) + 1.727 = 109.74$ (C_α—H), 109.18 (C_β—H), and 108.72 kcal/mol (C_γ—H). The sum $\Sigma\varepsilon_{ij}$ over all the bonds is 1975.30 kcal/mol, in good agreement with the value deduced from experimental results (1975.53 kcal/mol), and has been derived without the detailed knowledge of the distribution of charges among the individual hydrogen atoms. As regards the part due to the proper charge renormalization, $\Sigma_i\Sigma_j a_{ij}\Delta q_i = 1970.3 - \Sigma\varepsilon_{ij}^\circ = 41.9$ kcal/mol, it is clearly a major contribution to the energy of atomization of this molecule.

Bond-by-bond calculations are instructive in that they yield detailed pictures of the individual energy contributions. For homologous series, like the dialkylethers studied here, it is easy and convenient to derive general formulas for evaluating atomization energies. The calculation of the $\Sigma\varepsilon_{ij}^\circ$ energy being trivial, we focus attention on the $\Sigma_i\Sigma_j a_{ij}\Delta q_i$ part. For convenience we express the Δq_i's in terms of their NMR shifts and the a_{ij}'s in kcal mol^{-1}

ppm^{-1}, where appropriate. Noting that $q_{C\alpha} - q_C^{\circ} = q_{C\alpha} - q_{C\alpha}^{\circ} + q_{C\alpha}^{\circ} - q_C^{\circ}$, that is,

$$q_{C\alpha} - q_C^{\circ} = \Delta q_{C\alpha} + \Delta q_{C\alpha}^{\circ} \qquad (9.5)$$

is the Δq_C value to be used in calculations of CC and CH bond contributions involving α-carbons, straightforward summation yields[1].

$$\sum_i \sum_j a_{ij} \Delta q_i = \lambda_1 \sum N_{CC} \delta_C + \lambda_2 \sum \delta_{C \neq \alpha} + \lambda_3 \sum \delta_{C\alpha} + \lambda_4 \delta_O$$

$$\qquad (9.6)$$

$$+ (n_C - 2) a_{HC} q_H^{\circ} + \Delta q_{C\alpha}^{\circ} (a_{CC} \sum N_{CC}^{\alpha} + a_{CH} \sum N_{CH}^{\alpha} - 2a_{HC})$$

$$- a_{HC} q_O^{\circ}$$

where N_{CC} is the number of CC bonds formed by the carbon atom whose shift is δ_C. The $\Sigma N_{CC} \delta_C$ term includes both $\delta_{C\alpha}$ (relative to the diethylether α-carbon) and $\delta_{C \neq \alpha}$ (from ethane). N_{CC}^{α} and N_{CH}^{α} are, respectively, the number of CC and CH bonds formed by the α-carbons. The $\lambda_1 = a_{CC} - a_{CH} = 0.0356$ and $\lambda_2 = 4a_{CH} - a_{HC} = 0.0529$ (kcal mol^{-1} ppm^{-1}) parameters are those encountered earlier in Eqs. 6.3 and 6.4 (see also Chapter 6.4), while λ_3 and λ_4 are given by:

$$\lambda_3 = 3a_{CH} + a_{CO} - a_{HC}, \qquad (9.7)$$

$$\lambda_4 = 2a_{OC} - a_{HC}. \qquad (9.8)$$

Using the appropriate a_{ij} values (Table 5.7) and charge–shift relationships (Eqs. 4.10 and 4.11), it follows that $\lambda_3 = 0.1217$ kcal mol^{-1} ppm^{-1}. Because of relatively important variations in Δq_O, the $(1/2)(\partial^2 E_i^{vs}/\partial N_i^2)^{\circ} \Delta q_i$ term of

TABLE 9.1. ^{13}C and ^{17}O NMR Results (in ppm) for Ethers for Use in Eq. 9.6

Molecule	$\Sigma N_{CC}\delta_C$	$\Sigma \delta_{C \neq \alpha}$	$\Sigma \delta_{C\alpha}$	δ_O
$(CH_3)_2O$	0	0	-12.4	-59.0
$CH_3OC_2H_5$	10.7	8.9	-6.6	-29.0
$CH_3O\text{-}n\text{-}C_3H_7$	48.5	22.3	1.2	-35.0
$CH_3O\text{-}i\text{-}C_3H_7$	46.0	32.0	-3.3	-8.5
$(C_2H_5)_2O$	19.2	19.2	0	0
$(n\text{-}C_3H_7)_2O$	94.0	45.2	13.6	-10.0
$(i\text{-}C_3H_7)_2O$	78.4	68.8	4.8	46.0
$(n\text{-}C_4H_9)_2O$	187.6	97.2	9.6	-13.5
$(s\text{-}C_4H_9)_2O$	168.2	86.1	16.7	35.0
Tetrahydropyran		63.2	4.6	-6.0
1,4-Dioxane		7.6		-16.0

The $\delta_{C \neq \alpha}$ data (Refs. 2 and 4) are relative to ethane (5.8 ppm from TMS) and the $\delta_{C\alpha}$ shifts are from diethylether. The ^{17}O shifts (Ref. 2) are relative to diethylether. The $\Sigma N_{CC}\delta_C$ term includes both $\delta_{C\alpha}$ and $\delta_{C \neq \alpha}$. The tetrahydropyran and 1,4-dioxane ^{13}C data are taken from Stothers[4] and the ^{17}O results are extracted from Ref. 5. The results for the latter two compounds were deduced from bond-by-bond calculations because Eq. 9.6 is not applicable in these cases.

Eq. 5.47 must be considered in the evaluation of a_{OC}. In practice, it suffices to use $\lambda_4 = 0.1007$ kcal mol^{-1} ppm^{-1} for $a_{OC} = -0.804$ a.u. ($\delta_O < 0$) or 0.0994 kcal mol^{-1} ppm^{-1} for $a_{OC} = -0.800$ a.u. ($\delta_O > 0$). Equation 9.6 enables an empirical evaluation of $\Delta q^\circ_{C_\alpha}$. Indeed, the comparison of experimental atomization energies ΔE_a^* (Eq. 5.5), corrected for nonbonded contributions E_{nb}^* (Eqs. 5.30 and 5.31), with their theoretical counterparts deduced from Eq. 9.6 and the appropriate ε_{ij}°'s gives the estimate for $\Delta q^\circ_{C_\alpha}$ discussed previously (Eq. 9.3) and, therefrom, $q_O^\circ = 5.18$ me (Chapter 4.6). The appropriate shift values are indicated in Table 9.1.

The calculation of $\Sigma_i \Sigma_j a_{ij} \Delta q_i$ and of ΔE_a^{*bonds} (Eq. 5.48) is now straightforward. An estimate of the small nonbonded contributions is obtained from Eq. 5.30 and the approximations outlined in Chapter 4.6 for the atomic net charges. The theoretical atomization energies ΔE_a^* derived in this fashion are seen to agree within ~ 0.16 kcal/mol (average deviation) with their counterparts deduced from experimental results using Eq. 5.5 (Table 9.2).

While this sort of agreement prevents us from retracing minor uncertainties associated with our approximations, it also appears that reliable predictions can be made in a number of cases (Table 9.3).

The quality of calculated enthalpies rests, of course, largely with the approximations involved in the estimate of ZPE + H_T-H_0 energies. The use of vibrational energies constructed on the basis of regular group increments may be regarded acceptable in simple cases like those investigated here, as discussed in Chapter 7.7. It remains, however, that one should refrain from extrapolations in highly crowded systems until these are adequately supported by spectroscopic evidence. Moreover, the validity of the charge-shift correlations used in applications of Eq. 9.6 should also be scrutinized in these cases. With these reservations in mind, the predicted ΔH_f° values given in Table 9.3 should represent reasonably good estimates.

4. CARBONYL COMPOUNDS

The charge effects governing the energies of simple carbonyl compounds RCHO and RCOR' (R, R' = alkyl groups) are conveniently investigated by means of an appropriate general formula describing the $\Sigma_i \Sigma_j a_{ij} \Delta q_i$ term. The sum $\Sigma \Delta q_H$ follows from charge normalization (Eq. 9.1); q_C° and q_H° have their usual meaning (i.e., the C and H net charges of ethane, which is the reference compound for the alkyl part); $q_{C_\alpha}^\circ$ and q_O° are the carbonyl-C and O net charges, respectively, of acetone, whose CO bond energy (ε_{CO}°) is selected as reference. Of course, attention must be given to the $\Delta q_{C_\alpha}^\circ$ term (Eq. 9.2) because the appropriate reference charge for CO bond calculations is $q_{C_\alpha}^\circ$, whereas in all calculations of CC and CH bond energies, including those involving α-carbons, the reference charge is that of the ethane carbon atom, q_C° (see Eq. 9.5). With these premises, straightforward summation of the $a_{ij} \Delta q_i$ terms can be carried out, using also Eqs. 9.1 and 9.5. The following

TABLE 9.2. Comparison between Calculated and Experimental Energies of Atomization of Selected Ethers (kcal/mol)

Molecule	ΔH_f° (298.15, gas)	$ZPE + H_T\text{-}H_0$	$\Sigma \varepsilon_{ij}^\circ$	$\Sigma_i \Sigma_j a_{ij} \Delta q_i$	E_{nb}^*	ΔE_a^*	
						Calcd.	Exptl.
$(CH_3)_2O$	-43.99 ± 0.12	52.55	800.42	-3.33	-0.05	797.14	797.07
$CH_3OC_2H_5$	-51.72 ± 0.16	70.19	1083.67	9.56	-0.09	1093.32	1093.07
$CH_3O\text{-}n\text{-}C_3H_7$	-56.82 ± 0.26	87.83	1366.92	19.35	-0.25	1386.52	1386.43
$CH_3O\text{-}i\text{-}C_3H_7$	-60.24 ± 0.23	87.83	1366.92	22.82	-0.12	1389.86	1389.85
$(C_2H_5)_2O$	-60.26 ± 0.19	87.83	1366.92	22.45	-0.17	1389.54	1389.87
$(n\text{-}C_3H_7)_2O$	-69.85 ± 0.40	123.11	1933.42	41.92	-0.47	1975.81	1976.00
$(i\text{-}C_3H_7)_2O$	-76.20 ± 0.54	123.11	1933.42	48.97	-0.29	1982.68	1982.35
$(n\text{-}C_4H_9)_2O$	-79.82 ± 0.27	158.39	2499.92	61.95	-0.73	2562.60	2562.50
$(s\text{-}C_4H_9)_2O$	-86.26 ± 0.41	158.39	2499.92	68.22	-0.58	2568.72	2568.94
Tetrahydropyran	-53.39 ± 0.24	92.87	1506.18	50.95	-0.49	1557.62	1557.45
1,4-Dioxane	-75.51 ± 0.17	78.48	1312.86	39.59	-0.29	1352.74	1352.62

The ΔH_f° values are taken from J.D. Cox and G. Pilcher[6]. The $ZPE + H_T\text{-}H_0$ values are those of Table 7.6 (see Chapter 7.7).

TABLE 9.3. Predicted Enthalpies of Formation (298.15 K, gas) (kcal/mol)*

Molecule	$\Sigma N_{CC}\delta_C$	$\Sigma\delta_{C\neq\alpha}$	$\Sigma\delta_{C\alpha}$	δ_O	$\Sigma\varepsilon^\circ_{ij}$	$\Sigma_i\Sigma_j a_{ij}\Delta q_i$	E^*_{nb}	ΔE^*_a	ΔH°_f
CH_3O-n-C_4H_9	95.4	48.4	-0.8	-35.0	1650.17	29.55	-0.37	1680.09	-62.21
CH_3O-i-C_4H_9	110.0	50.1	6.7	-36.5	1650.17	30.93	-0.34	1681.44	-63.56
CH_3O-s-C_4H_9	88.0	40.2	2.2	-15.0	1650.17	32.16	-0.28	1682.61	-64.73
CH_3O-n-C_5H_{11}	142.3	71.7	-0.5	-34.0	1933.42	39.99	-0.49	1973.90	-67.75
CH_3O-i-C_5H_{11}	163.7	86.3	-2.2	-36.0	1933.42	41.11	-0.45	1974.98	-68.83
CH_3O-s-C_5H_{11}	122.6	47.4	8.0	-23.0	1933.42	41.06	-0.39	1974.87	-68.72
C_2H_5O-n-C_3H_7	56.2	31.9	6.8	-4.8	1650.17	32.17	-0.32	1682.66	-64.78
C_2H_5O-i-C_3H_7	50.7	42.9	2.6	21.5	1650.17	35.59	-0.22	1685.98	-68.10
C_2H_5O-n-C_4H_9	103.5	58.2	4.9	-8.0	1933.42	42.09	-0.44	1975.95	-69.80
C_2H_5O-i-C_4H_9	117.7	59.7	12.2	-7.5	1933.42	43.61	-0.41	1977.44	-71.29
C_2H_5O-s-C_4H_9	94.3	51.3	8.5	18.0	1933.42	45.35	-0.35	1979.12	-72.97
n-C_3H_7O-i-C_3H_7	88.7	55.9	9.4	17.5	1933.42	45.46	-0.37	1979.25	-73.10

*The estimated ZPE + H_T-H_0 energies are 105.47 and 123.11 kcal/mol, respectively, for the $C_5H_{12}O$ and $C_6H_{14}O$ ethers, using the approximation indicated in Table 7.6. The NMR data are taken from Ref. 2.

result is obtained[1], in which the Δq_i's are expressed in terms of the corresponding NMR shift values:

$$\sum_i \sum_j a_{ij} \Delta q_i = \lambda_1 \sum N_{CC} \delta_C + \lambda_2 \sum \delta_{C \neq \alpha} + \lambda_3' \delta_{C\alpha} + \lambda_4' \delta_O + n_C a_{HC} q_H^\circ$$

$$+ \Delta q_{C\alpha}^\circ (a_{C\alpha C} N_{CC}^\alpha + a_{C\alpha H} N_{CH}^\alpha - a_{HC}) - a_{HC} q_O^\circ, \tag{9.9}$$

where $\lambda_1 = 0.0356$ and $\lambda_2 = 0.0529$ kcal mol^{-1} ppm^{-1} have their usual meaning (Eqs. 6.3 and 6.4), as does N_{CC} (i.e., the number of CC bonds formed by the C atom whose shift is δ_C; the $\Sigma N_{CC} \delta_C$ term includes both $\delta_{C\alpha}$ and $\delta_{C \neq \alpha}$.); N_{CC}^α and N_{CH}^α are, respectively, the number of CC and CH bonds formed by the carbonyl-C atom, and

$$\lambda_3' = 2a_{C\alpha H} + a_{CO} - a_{HC} \tag{9.10}$$

$$\lambda_4' = a_{OC} - a_{HC} \tag{9.11}$$

While λ_1 and λ_2 are already well-known from the study of the saturated hydrocarbons, meaning that the calculation of $\lambda_1 \Sigma N_{CC} \delta_C + \lambda_2 \Sigma \delta_{C \neq \alpha}$ presents no difficulty, there are problems with the evaluation of λ_3' and λ_4'. To begin with, it is difficult to assess which is the most appropriate value of the derivative $(\partial E_i^{vs} / \partial N_i)^\circ$ for carbonyl carbon atoms, which enters the calculation of $a_{C\alpha C}$, $a_{C\alpha H}$, and a_{CO} through Eq. 5.47. This uncertainty, however, does not affect λ_1 because the $a_{C\alpha C} - a_{C\alpha H} = a_{CC} - a_{CH}$ term appearing in this calculation does not depend on $(\partial E_i^{vs} / \partial N_i)^\circ$. The selected a_{ij} values (Table 5.7) are plausible on grounds that at carbonyl-carbon atoms mainly $2p$ populations are affected by substitution. The fact that the selected value

$$\lambda_3' = 0.0675 \text{ kcal mol}^{-1} \text{ ppm}^{-1}$$

seems appropriate in actual calculations using Eq. 9.9 is not a convincing argument because (i) this estimate involves the acceptance of the C-charge–shift relationship, Eq. 4.10, which is not sufficiently substantiated for

TABLE 9.4. ^{13}C and ^{17}O NMR Results (in ppm) for Selected Aldehydes and Ketones

Molecule	$\Sigma N_{CC} \delta_C$	$\Sigma \delta_{C \neq \alpha}$	$\delta_{C\alpha}$	δ_O
CH_3CHO	19.3	25.3	-5.5	23.0
C_2H_5CHO	57.6	30.1	-3.3	10.5
n-C_3H_7CHO	103.1	57.0	-3.5	20.0
$(CH_3)_2CO$	44.4	44.4	0	0
$CH_3COC_2H_5$	84.9	52.9	1.2	-11.5
CH_3CO-n-C_3H_7	135.8	81.9	1.5	-6.0
CH_3CO-i-C_3H_7	158.8	81.0	4.0	-12.0
$(C_2H_5)_2CO$	128.8	61.6	4.2	-22.0
CH_3CO-n-C_4H_9	198.6	113.1	2.0	-6.0
$(i$-$C_3H_7)_2CO$	264.2	113.6	10.5	-34.0

The $\delta_{C \neq \alpha}$ data[4] are relative to ethane (5.8 ppm from TMS) and the $\delta_{C\alpha}$ shifts[4] are from acetone. The ^{17}O shifts[3] are relative to acetone. The $\Sigma N_{CC} \delta_C$ term includes both $\delta_{C\alpha}$ and $\delta_{C \neq \alpha}$.

TABLE 9.5. Charge–Dependent Energy Contributions of Selected Carbonyl Compounds (kcal/mol)

Molecule	$\lambda_1 \Sigma N_{CC}$ $+\lambda_2 \Sigma \delta_{C \neq \alpha}$	$\lambda'_3 \delta_{C\alpha}$	$\lambda'_4 \delta_O$	a_{OC}(a.u.)	λ'_4(a.u.)
CH_3CHO	2.04	−0.37	−2.87	−1.081	−0.074
C_2H_5CHO	3.64	−0.22	−1.16	−1.072	−0.065
$n\text{-}C_3H_7CHO$	6.69	−0.24	−2.42	−1.079	−0.072
$(CH_3)_2CO$	3.93	0	0	−1.065	−0.058
$CH_3COC_2H_5$	5.82	0.08	0.97	−1.057	−0.050
$CH_3CO\text{-}n\text{-}C_3H_7$	9.17	0.10	0.55	−1.061	−0.054
$CH_3CO\text{-}i\text{-}C_3H_7$	9.94	0.27	1.01	−1.057	−0.050
$(C_2H_5)_2CO$	7.84	0.28	1.59	−1.050	−0.043
$CH_3CO\text{-}n\text{-}C_4H_9$	13.05	0.14	0.55	−1.061	−0.054
$(i\text{-}C_3H_7)_2CO$	15.41	0.71	1.98	−1.041	−0.034

carbonyl carbons, and (ii) because the range of variation of the $\delta_{C\alpha}$ shifts is relatively small (Table 9.4). On the other hand, however, a possible error associated with the $(\partial E_i^{vs}/\partial N_i)°$ derivative would affect the $\Delta q_{C\alpha}°(a_{C\alpha C}N_{CC}^\alpha + a_{C\alpha H}N_{CH}^\alpha - a_{HC})$ term of Eq. 9.9 and result in distortions in empirical estimates of $\Delta q_{C\alpha}°$ and $\varepsilon_{CO}°$ from experimental ΔE_a^* energies. Now, both the $\Delta q_{C\alpha}°$ (Eq. 9.4) and the $\varepsilon_{CO}°$ value (Table 5.6) obtained from a fit with experimental ΔE_a^* results are in reasonable agreement with theoretical expectations, suggesting that any anticipated error in $(\partial E_i^{vs}/\partial N_i)°$ is probably not too severe. The smallness of the $\lambda'_3 \delta_{C\alpha}$ contributions, as compared with the $\lambda_1 \Sigma N_{CC} \delta_C + \lambda_2 \Sigma \delta_{C \neq \alpha}$ terms, is illustrated in Table 9.5.

As regards the carbonyl oxygen atoms, ab initio calculations indicate (i) an upfield shift with increasing electron population and (ii) the predominant role of $2p$ electrons in governing the variations in net atomic charges. Beyond these qualitative results, however, no clear charge–shift correlation was established, mainly because the charge results appear to be particularly affected by any imperfection in attempted optimizations of the ζ exponents. In contrast with other ^{13}C and ^{17}O charge–shift correlations which are based on theoretical electron populations, the relationship

$$\Delta q_O \text{ (carbonyl)} = 2.7 \delta_O \text{ (me)} \qquad (9.12)$$

is deduced from, and justified by, a numerical analysis involving Eq. 9.9 and experimental energies: while not ranking at the same degree of confidence as the other charge–shift correlations which were used so far, Eq. 9.12 appears to represent at least a reasonable approximation. The true merits of Eq. 9.12 are best recognized in the fact that the 2.7 factor regulates the value of λ'_4 (expressed in kcal mol^{-1} ppm^{-1}) in two ways. Firstly, it appears in the unit conversion factor which is given in the following equation:

$$\lambda'_4 \text{ (in atomic units) times } \frac{627.51}{1000} \times 2.7 = \lambda'_4 \text{ in kcal mol}^{-1} \text{ ppm}^{-1} \text{ units.}$$

Secondly, it determines the value of the $(1/2)(\partial^2 E_i^{vs}/\partial N_i^2)^\circ \Delta q_i$ term appearing in a_{OC} (Eq. 5.47). This correction is particularly important for carbonyl oxygen atoms, as indicated by the a_{OC} results given in Table 9.5. The effect of this correction is further magnified in the λ_4' parameter (Eq. 9.11) which, in contrast with the other λ parameters, is far from being constant (Table 9.5). The 2.7 coefficient of Eq. 9.12 is thus subject to a double constraint, i.e., it must not only simple adjust $\lambda_4'\delta_O$ to give the correct final sum in the expression for $\Sigma_i\Sigma_j a_{ij}\Delta q_i$ (Eq. 9.9) but also must produce the correct variations of λ_4' via the $(\partial^2 E_i^{vs}/\partial N_i^2)^\circ$ derivative which is given by theory. Although the $\Sigma_i\Sigma_j a_{ij}\Delta q_i$ sums are largely controlled by all the other terms of Eq. 9.9, namely by the $\lambda_1 \Sigma N_{CC}\delta_C + \lambda_2\Sigma\delta_{C\neq\alpha}$ part, it is clear that there is not much room left for error in the $\lambda_4'\delta_O$ part if satisfactory results are to be obtained. The quality of the calculations carried out in this manner is best illustrated by comparisons with experimental results (Table 9.6).

These comparisons are certainly satisfactory because the average deviation (0.16 kcal/mol) between calculated and experimental enthalpies of formation is well within experimental uncertainties. There are, of course, some reservations to be made as regards the nonbonded contributions derived from Eq. 5.30, because it is presently difficult to assess how valid their estimates are. Moreover, some (probably minor) uncertainties are associated with the estimates of the $ZPE + H_T-H_0$ energies. However, considering now the predominant role played by those terms of Eq. 9.9 which are best known (namely, the sum $\lambda_1\Sigma N_{CC}\delta_C + \lambda_2\Sigma\delta_{C\neq\alpha}$ which accounts for ~ 80–95% of all the parts depending on λ parameters), it seems fair to say that the results obtained for the carbonyl compounds support the general validity of the theory underlying these calculations.

Although it is understandably always pleasing to witness satisfactory comparisons between theory and experiment, one should not evaluate the merits of an approach solely by its ability to reproduce experimental results. Useful insights may be provided by general qualitative views suggested by the present results, as indicated by the examples given below.

5. CONCLUSIONS

The best insight into the charge-related fine-tuning of molecular energies is offered by isomers in which the number of the individual types of bonds (and, hence, $\Sigma\varepsilon_{ij}^\circ$) are the same. The charge effects accompanying isodesmic structural rearrangements are, for the largest part, contained in the $\Sigma_i\Sigma_j a_{ij}\Delta q_i$ terms relating to the changes in the bonded contributions, whereas the changes in nonbonded interactions are only minor ones. For example, in going from $(n\text{-}C_3H_7)_2O$ to $(i\text{-}C_3H_7)_2O$ the chemical bonds gain 7.05 kcal/mol in stability, whereas the nonbonded part destabilizes the branched isomer by 0.18 kcal/mol (Table 9.2). Similarly, the isodesmic structural change $C_2H_5CHO \rightarrow (CH_3)_2CO$ is accompanied by a stabilization of 6.68 kcal/mol

TABLE 9.6. Comparison between Calculated and Experimental Atomization Energies of Selected Carbonyl Compounds (kcal/mol)

Molecule	ΔH_f° (298.15, gas)	$ZPE + H_T\text{-}H_0$	$\Sigma \varepsilon_{ij}^\circ$	$\Sigma_i \Sigma_j a_{ij} \Delta q_i$	E_{nb}^*	ΔE_a^* calcd.	ΔE_a^* exptl.
CH_3CHO	-39.73 ± 0.12	36.65	676.25	-0.78	-0.30	675.77	675.69
C_2H_5CHO	-45.45 ± 0.21	54.95	959.50	10.17	-0.36	970.03	970.34
$n\text{-}C_3H_7CHO$	-48.98 ± 0.34	73.25	1242.75	19.26	-0.75	1262.76	1262.76
$(CH_3)_2CO$	-51.90 ± 0.12	54.59	959.50	16.85	-0.07	976.42	976.43
$CH_3COC_2H_5$	-57.02 ± 0.20	72.89	1242.75	27.19	-0.45	1270.39	1270.48
$CH_3CO\text{-}n\text{-}C_3H_7$	-61.92 ± 0.26	91.19	1526.00	37.52	-0.42	1563.94	1564.30
$CH_3CO\text{-}i\text{-}C_3H_7$	-62.76 ± 0.21	91.19	1526.00	38.75	-0.55	1565.30	1565.14
$(C_2H_5)_2CO$	-61.65 ± 0.21	91.52	1526.00	37.42	-0.92	1564.34	1564.36
$CH_3CO\text{-}n\text{-}C_4H_9$	-66.96 ± 0.28	109.49	1809.25	48.82	-0.55	1858.62	1858.27
$(i\text{-}C_3H_7)_2CO$	-74.40 ± 0.28	128.12	2092.50	60.60	-1.85	2154.95	2154.97

The ΔH_f° values are extracted from Ref. 6. As regards the $ZPE + H_T\text{-}H_0$ energies, an increment of 18.3 kcal/mol was assumed for each added CH_2 group with respect to the closest "parent" compound, in line with the results obtained for the parent hydrocarbons[7]; see also Table 7.6.

in the bonded part, and a destabilization of 0.29 kcal/mol due to the change in the interactions between nonbonded atoms (Table 9.6). The general conclusion is that any significant energy difference between isomers is mainly due to charge effects governing the energies of the chemical bonds.

It appears interesting to inquire about the general features of the intra-molecular charge redistributions which take place during structural rear-rangements. With the saturated hydrocarbons, the picture was simple: the loss of electronic charge at the hydrogen atoms during an isodesmic structural change resulting in a more stable molecule can obviously occur only in favor of the carbon atoms, a process which we describe as a H → C intramolecular charge transfer. With molecules containing heteroatoms, however, the situa-tion is different because other types of charge transfers, e.g., H → O, need also be considered. Pertinent conclusions regarding this matter can be drawn from the results presented for the ethers and the carbonyl compounds. Using the chemical shifts measured by Delseth and Kintzinger[2,3] (most of which are reported in Tables 9.1, 9.3, and 9.4) and the appropriate charge-shift relationships, it is easy to calculate $\Sigma\Delta q_C$ and Δq_O. Therefrom, using now Eq. 9.1, the sum $\Sigma\Delta q_H$ is also readily deduced. In this fashion it appears, for example, that the change C_2H_5O-n-$C_4H_9 \rightarrow C_2H_5O$-$s$-$C_4H_9$ is accompanied by a total loss of 6.45 me at the hydrogen atoms and a gain of 6.94 me at the oxygen atom, including some electronic charge (0.49 me) lost by the carbon atoms. The main cause accounting for the gain $\Delta(\Delta E_a^*) = 3.17$ kcal/mol in stabilization is thus a H → O electron transfer. The results for the ethers are collected in Table 9.7 and suggest the following conclusions.

For structural changes affecting α-carbons (**1–9**), important stabilizations (say, $\geqslant 2$ kcal/mol) result mainly from a H → O electron transfer with little change in the carbon populations (**1, 2, 4, 5, 7–9**). When the H → O charge transfer is small (**3, 6**), only a minor gain in stability is achieved, resulting in essence from a C → O charge transfer. Structural changes involving β-carbons (**10–13**), on the other hand, result in a stabilization of the alkyl part, just as in saturated hydrocarbons, with only a minor participation of the oxygen charges. The main driving factor is then a H → C charge reorganization. The $\Delta(\Delta E_a^*)$'s are accordingly smaller, resembling those observed for the saturated hydrocarbons (e.g., n-$C_4H_{10} \rightarrow i$-C_4H_{10} with $\Delta(\Delta E_a^*) = 1.92$ or n-$C_5H_{12} \rightarrow i$-C_5H_{12} with $\Delta(\Delta E_a^*) = 1.27$ kcal/mol, from Table 6.1), con-trasting with isomerizations involving α-carbons where the change from n-C_3H_7 to i-C_3H_7 represents $\Delta(\Delta E_a^*) \simeq 3.4$ kcal/mol (**1, 4, 7, 8**). In short, *the main driving forces are* H → O *charge transfers when significant stabiliza-tions are achieved in structural changes involving α-carbons, or* H → C *charge transfers when the stabilization occurs mainly in the alkyl part as a result of structural changes involving carbon atoms which are sufficiently distant from the oxygen atom.* The role of the alkyl hydrogen atoms acting as "electron reservoirs" is clearly illustrated by these results.

A similar conclusion can also be drawn for the carbonyl compounds. Indeed, the structural change $C_2H_5CHO \rightarrow (CH_3)_2CO$ [$\Delta(\Delta E_a^*) = 6.39$

TABLE 9.7. Charge Effects Accompanying Isodesmic Structural Changes in Dialkylethers

Isomerization	$\Delta(\Delta E_a^*)$ kcal/mol	Change in net charge (me) at the		
		H atoms	C atoms	O atoms
1 $CH_3O\text{-}n\text{-}C_3H_7$ → $CH_3O\text{-}i\text{-}C_3H_7$	3.34	7.85	−0.77	−7.08
2 $CH_3O\text{-}n\text{-}C_4H_9$ → $CH_3O\text{-}s\text{-}C_4H_9$	2.52	4.57	0.77	−5.34
3 $CH_3O\text{-}n\text{-}C_5H_{11}$ → $CH_3O\text{-}s\text{-}C_5H_{11}$	0.97	0.60	2.34	−2.94
4 $C_2H_5O\text{-}n\text{-}C_3H_7$ → $C_2H_5O\text{-}i\text{-}C_3H_7$	3.32	8.03	−1.01	−7.02
5 $C_2H_5O\text{-}n\text{-}C_4H_9$ → $C_2H_5O\text{-}s\text{-}C_4H_9$	3.17	6.45	0.49	−6.94
6 $C_2H_5O\text{-}n\text{-}C_5H_{11}$ → $C_2H_5O\text{-}s\text{-}C_5H_{11}$	0.40	1.55	2.06	−3.61
7 $n\text{-}C_3H_7O\text{-}n\text{-}C_3H_7$ → $n\text{-}C_3H_7O\text{-}i\text{-}C_3H_7$	3.44	8.30	−0.96	−7.34
8 $n\text{-}C_3H_7O\text{-}i\text{-}C_3H_7$ → $i\text{-}C_3H_7O\text{-}i\text{-}C_3H_7$	3.43	8.84	−1.23	−7.61
9 $n\text{-}C_4H_9O\text{-}n\text{-}C_4H_9$ → $s\text{-}C_4H_9O\text{-}s\text{-}C_4H_9$	6.12	12.36	0.59	−12.95
10 $CH_3O\text{-}n\text{-}C_4H_9$ → $CH_3O\text{-}i\text{-}C_4H_9$	1.35	0.96	−1.36	0.40
11 $CH_3O\text{-}n\text{-}C_5H_{11}$ → $CH_3O\text{-}i\text{-}C_5H_{11}$	1.08	1.38	−1.91	0.53
12 $C_2H_5O\text{-}n\text{-}C_4H_9$ → $C_2H_5O\text{-}i\text{-}C_4H_5$	1.49	1.43	−1.30	−0.13
13 $C_2H_5O\text{-}n\text{-}C_5H_{11}$ → $C_2H_5O\text{-}neo\text{-}C_5H_{11}$	3.15	2.82	−4.16	1.34

A positive value for the change in net charge represents a loss of electronic charge.

kcal/mol] involves a gain of 28.35 me at the oxygen and of 2.55 me at the carbon atoms, at the expense of the electron populations at the hydrogen atoms. The change n-$C_3H_7CHO \rightarrow CH_3COC_2H_5$ [$\Delta(\Delta E_a^*) = 7.63$ kcal/mol] is accompanied by a gain of 85.05 me at the oxygen atom and a minimal gain (0.09 me) at the carbon atoms, as a result of an important loss of electronic charge at the hydrogen atoms. Finally, the change CH_3CO-n-$C_3H_7 \rightarrow CH_3CO$-$i$-$C_3H_7$ [$\Delta(\Delta E_a^*) = 1.36$ kcal/mol] is also accompanied by a major gain at the oxygen atom (16.20 me) and a minor one (0.24 me) at the carbon atoms, at the expense of the electron populations at the hydrogen atoms. Briefly, the major charge transfer in the isodesmic transformations involving these carbonyl compounds is $H \rightarrow O$ and is significantly more important (in terms of electron redistributions) than that occurring in the series of dialkylethers. Although the charge transfer values in carbonyl compounds cannot be assessed with the same degree of confidence as for the hydrocarbons and ethers, it remains that the general trends indicated here are expected to be essentially correct.

Finally, it is noteworthy that the isodesmic structural changes considered here occur with virtually no change in vibrational energy, the largest variations (0.36 kcal/mol) being those for the aldehyde $\rightarrow$ ketone transformations. While this statement should not be unduly generalized, particularly to highly crowded systems, it is fair to say that the thermophysical molecular stabilities of these molecules at, say, 25°C, are largely governed by the structure-related charge effects described by the $\Sigma_i\Sigma_j a_{ij}\Delta q_i$ term.

REFERENCES

1. S. Fliszár and M.-T. Béraldin, *Can. J. Chem.*, **60**, 792 (1982).
2. C. Delseth and J.P. Kintzinger, *Helv. Chim. Acta*, **61**, 1327 (1978).
3. C. Delseth and J.P. Kintzinger, *Helv. Chim. Acta*, **59**, 466 (1976); *ibid.*, **59**, 1411 (1976).
4. J.B. Stothers, "Carbon–13 NMR Spectroscopy", Academic Press, New York, NY, 1972.
5. T. Sugawara, Y. Kawada, M. Katoh, and H. Iwamura, *Bull. Chem. Soc. Japan*, **52**, 3391 (1979).
6. J.D. Cox and G. Pilcher, "Thermochemistry of Organic and Organometallic Compounds", Academic Press, New York, NY, 1970.
7. S. Fliszár and J.-L. Cantara, *Can J. Chem.*, **59**, 1381 (1981).

CHAPTER 10

Conclusion and Assessment

1. CHARGE DISTRIBUTIONS

The physically meaningful partitioning of the electronic charge of a molecule among its individual atoms, briefly, the appropriate definition of atomic charges, is a problem laden with difficulties. A convenient way of extracting atomic charges is offered by Mulliken's population analysis, which is rooted in the LCAO formalism. However, a superficial inspection of numerous charge results deduced from a variety of LCAO wave functions (both semi-empirical and *ab initio*) seems to discredit, at one point or another, at least an important part of the methods applied. Evidently, it makes no sense to deduce a positive net charge of ~ 0.04 electron or a negative charge of -1.07 e for the carbon atom of methane, or any value between (or outside) these extremes, depending upon the method of calculation, yet this is precisely the situation which is encountered in current charge analyses based on Mulliken's scheme. Understandably, this state of affairs casts serious doubts on the ability of the Mulliken scheme to obtain meaningful charge distributions, suited for the discussion of real physical problems; moreover, any theoretical improvement in the basis set description in *ab initio* calculations only worsens the charge results, leading eventually to entirely unrealistic charge separations (like the electron population of ~ 7.07 e on the methane-C atom).

Fortunately, the picture regarding Mulliken charges is not quite as bad as a superficial inspection of the raw data would seem to indicate: one should not only brandish the difficulties outlined above but also consider the positive aspects contained in this type of charge analysis. To begin with, it is

TABLE 10.1. Net Charges of the Propane Methyl Group and of its Carbon and Hydrogen Atoms (in 10^{-3} electron units) Deduced from Selected Theoretical Methods

Method	Net Charge (me)		
	CH_3 group	Methyl-C	Methyl-H
INDO	-8	67	-25
STO–3G	-5.12	-23.81	6.23
GTO($6s3p/3s$)	-8.8	-236.5	75.9
BO	-7	-508	167

most instructive to examine the net charge carried by a methyl group in propane and to compare the results deduced from different theoretical methods, as indicated in Table 10.1.

These examples show (*i*) that the various methods consistently predict that the methyl groups in propane are negatively charged, in agreement with experimental evidence[1], and (*ii*) that the results deduced from the different methods are relatively similar to one another, sharply contrasting in this respect with the results deduced for the individual C and H net charges of the methyl groups. The latter remark applies also to the charges of, and within, methyl groups attached to other alkyl groups, e.g., the isopropyl or the *tert*-butyl groups in isobutane and neopentane, respectively. It turns out that each theoretical method consistently predicts that a methyl group carries more electrons as the group to which it is attached is a better electron-donor, in the usual inductive order, i.e., *tert*-butyl $> \cdots >$ isopropyl $> \cdots >$ ethyl $>$ methyl. Moreover, each theoretical method predicts that a hydrogen atom carries more electronic charge as the alkyl group to which it is linked is a better electron-donor, in precisely the same order. The strong point of similarity existing between the sets of charge results deduced from different theoretical approaches resides in the fact that they all reflect the detailed facets of the familiar inductive effects in spite of the widely differing individual Mulliken atomic charges, depending upon the method of calculation. It remains, of course, that the major discrepancies between the charges deduced from the various methods are most disturbing.

These discrepancies are easily resolved, namely by assuming that the assignment of the overlap population terms need not necessarily follow the original Mulliken half-and-half partitioning scheme in cases involving heteronuclear overlap partners. This modification, coupled with the idea that the changes in electron populations occur "most reluctantly", thus keeping them as small as possible, yields new sets of atomic charges which are closely linked to, and easily deduced from, the original Mulliken populations. When derived from different wave functions, these modified atomic charges are not nearly as dissimilar as their original Mulliken counterparts. More significantly, the relative ordering of the carbon charges deduced in this fashion for an important number of compounds is the only one which corresponds to the ordering determined from comparisons with experimental quantities, namely ^{13}C nuclear magnetic resonance shifts, energies of atomization, and ionization potentials of alkanes[2]. From here on, charge-related chemical effects are discussed only in terms of net charges deduced from the "generalized Mulliken scheme" involving the appropriately modified partitioning of overlap populations.

To begin with, it appears that the familiar inductive effects of alkyl groups, commonly rationalized in terms of polar substituent constants, are clearly reflected in the electron distributions of alkanes. [Incidentally, the selection of the scale defining polar effects (σ^* or σ_I, following Taft[3]) is of no special concern because of the way these scales are related to one another[4].] The interpretation of these effects is rooted in an important result, namely the C^+—H^- polarity in saturated hydrocarbon systems. The idea suggested by this polarity is that hydrogen atoms "retain" electrons in CH_x groups. Charge analyses indicate, indeed, that this electron retention is more important in hydrogen-richer than in hydrogen-poorer CH_x groups, thus promoting intramolecular charge transfers in the direction $CH_2 \rightarrow CH_3$, $CH \rightarrow CH_2$, etc. As a rule of thumb, it appears that a quaternary carbon "loses" about three times as much, and a CH group about twice as much electronic charge to each adjacent CH_3 group as a CH_2 group. As a consequence, any methyl group in isobutane is electron-richer than in propane, the largest negative charge being carried by the neopentane-CH_3 groups. This, of course, is tantamount to saying that the electron-releasing abilities of the alkyl groups are in the order *tert*-butyl > isopropyl > ethyl > methyl. Similarly, referring now to isopentane, it follows that the methyl groups attached to CH are electron-richer than the methyl group linked to CH_2 (which "retains" electrons more efficiently than CH), meaning that a *sec*-butyl group is a better electron-donor than the isobutyl group. The rule regarding the increased "electron-retention" in hydrogen-richer CH_x groups applies quite generally, namely to hydrogen and oxygen atoms attached to CH_x. For example, a hydrogen is more negatively charged in $(CH_3)_3C$—H than in $(CH_3)_2CH$—H or CH_3CH_2—H. For the same reason, an oxygen atom attached to CH groups (as in diisopropyl ether) is more negative than the oxygen of di-*n*-propyl ether, and the electronic charge carried by the oxygen atom is larger in acetone than in ethanal. While these deductions are, in part, based on the fact that the individual C and H charges vary only moderately from case to case, it remains that relatively minor, but significant, structure-dependent variations in atomic charges do occur. The effect of α-methyl substitution is best illustrated using isobutane as an example. The apparent conflict between the facts (*i*) that a "central" carbon atom gains electrons with increasing number of methyl substituents and (*ii*) that the methyl groups withdraw electronic charge from the "central" CH_x group is easily resolved by observing that although each methyl group pulls electrons from the *iso*-C_3H_7 group [$q(CH_3) = -5.04$ me], it attracts less charge than the hydrogen atom which it has replaced [q_H(sec) in propane $=$ -14.09 me]. This, of course, means that hydrogen is a better attractor than CH_3, which is reflected in methane, $(CH_3)^+$—H^-, i.e., in the C^+—H^- polarity. In fact, the difference $-14.09 - (-5.04) = -9.05$ me represents the correct H net charge in methane. Similar arguments apply also to other molecules. Quite generally, the central carbon atom becomes increasingly negative (i.e., less positive) as the number of α-methyl substituents is increased, not because methyl itself pushes electrons toward the central carbon atom but because it withdraws less electrons than the hydrogen atom for which it has been substituted. It must be borne in mind, however, that not only the number, but also the quality, of α-substituents govern the local charge of a carbon atom. In neopentane, for example, the primary carbon atoms are less positive than the quaternary central atom because of the particularly strong electron-releasing ability of the *tert*-butyl group. Similarly, the secondary carbons of adamantane are less positive than the tertiary ones because, in a first approximation, their behavior can be considered similar to that of a CH_2 group attached to 2 isopropyl groups as compared to a CH group linked to 3 ethyl groups, which are less good donors than 2 isopropyl groups.

So far, we have learned (*i*) that atomic charges suited for the study of physical properties can be derived from a "generalized" Mulliken scheme with appropriate partitioning of overlap populations and (*ii*) that the C and H net charges obtained with these premises are "almost" invariant in saturated hydrocarbons and exhibit a C^+—H^- polarity. This defines the "electron retention" in CH_x groups in the order $CH_3 > CH_2 > CH > C$ and offers a qualitative account for the relative ordering of the inductive effects of alkyl groups. The bottom line is that the interpretation of the familiar inductive effects in terms of electron release is rooted in the C^+—H^- polarity, i.e., in a picture viewing the hydrogen atoms as "electron reservoirs". Quite significantly, this picture plays a prevailing role in the interpretation of the energetic effects accompanying structure-related electron distributions, namely in the evaluation of energy differences between isomers.

The general conclusions drawn for the alkyl groups in ethylenic and oxygen-containing compounds are quite similar to those deduced from the detailed analysis of paraffins. Relationships between theoretical charges and inductive effects are, in fact, known for a number of more or less complex organic systems, e.g., the carbon charge in a benzene ring in terms of the polar character of the *para* substituent[5]. It remains that the single most important difficulty encountered in the study of molecular energies in terms of electronic charge distributions rests precisely with the accurate calculation of atomic charges. Presently, the calculations are still somewhat involved because, in a number of cases, we have to resort to normalization procedures in order to make up for the lack of a comprehensive knowledge of atomic charges. In turn, the use of appropriate general formulas, in which hydrogen charges do not appear explicitly, makes it possible to describe large classes of compounds in terms of atoms other than hydrogen, namely C and O atoms, whose charges can be extracted from accurate empirical correlations with NMR shifts. While lacking the elegance of a purely theoretical approach, the judicious use of carefully established charge-NMR shift correlations fulfills its goals, although the significance or generality of such correlations is presently not known. Fortunately, encouraging theoretical progress is being made in the area of charge calculations: since it became clear that the CH polarity in simple alkanes is C^+—H^- (at variance with most results derived from Mulliken population analyses), new powerful and realistic methods[6] are converging toward this type of result, which is part of the present charge analyses. Hopefully, these methods will attain an adequate degree of accuracy and assist future charge calculations. It is also clear, however, that the extension of charge–shift correlations to other classes of molecules (alcohols, acetals, etc.) can be envisaged as a powerful means of studying large molecules of biological interest, which would normally lie outside the range of computational feasibility.

2. MOLECULAR ENERGIES

A molecule in its hypothetical vibrationless state can be considered as a collection of atoms with energies differing from their free-state values. The corresponding energy differences are, for each atom, a measure for the process of becoming part of a molecule and contain a portion of the molecular binding energy. Alternatively, in a mathematically equivalent description, molecules can be regarded as assemblies of chemical bonds—a model which retains the familiar structural formulas as a valuable basis for discussion. The latter description implies a separate evaluation of interaction energies between nonbonded atoms, which can be carried out with the help of Del Re's approximation

$$E(\text{nonbonded}) \simeq \frac{1}{2} \sum_{k,l} \frac{q_k q_l}{r_{kl}}$$

reflecting Coulomb interactions between net charges of nonbonded atoms k and l at a distance r_{kl}. The important point is that these nonbonded interactions cannot be made responsible for the bulk of the energy differences between structural isomers, since branching in alkyl chains destabilizes a molecule as far as nonbonded interactions are concerned, which is contrary to the experimental trend for total energies. For example, the nonbonded energy favors normal pentane *vs.* neopentane (by ~ 0.1 kcal/mol) although the latter is the more stable isomer (by ~ 4 kcal/mol). Similarly, di-*n*-propylether and diisopropylether differ from one another by ~ 0.2 kcal/mol in nonbonded energy in favor of the normal isomer, although the branched form is more stable, by ~ 6.8 kcal/mol. It is, therefore, the "bonded part" which primarily governs the structure-related energy effects. This "bonded part" is conveniently expressed in terms of individual energy contributions ε_{ij} associated with bonded atom pairs ij.

This result indicates that bond energy contributions (i.e., the ε_{ij}'s) cannot be simply transferred from one molecule to another. The physical reason can be visualized as follows, by considering that the transfer of an ij bond means, in fact, transferring the pair of bond-forming atoms i and j exactly as they are (including, of course, their electron clouds), if the bond energy is to remain unaltered. Yet, if constant electron populations $N_A(\neq Z_A)$, $N_B(\neq Z_B), \ldots$ are associated with all individual atoms A, B, $\ldots$ of an electroneutral molecule, any other non-isomeric molecule constructed from the same atoms with the same charges would not satisfy the requirement for molecular electroneutrality. For example, if the carbon and hydrogen net charges of ethane are 0.0351 and -0.0117 e, respectively, the methane "molecule" constructed from the same "atoms" would carry an excess electronic charge of -0.0117 electron. Hence, unless we deny the existence of intramolecular charge transfers, we cannot construct molecules with constant, exactly transferable, bond energy contributions. The obvious conclusion is that the solution of the energy problem for electroneutral molecules involves a consideration of the way individual bond energy contributions depend on local charges.

This approach is conveniently developed in two steps. First, we select an appropriate set of "reference bonds" with energies ε_{ij}° and reference atomic charges q_i° and q_j°. Next, we evaluate in what manner ε_{ij} differs from ε_{ij}° when the electron populations of the bond-forming atoms are no longer those of the corresponding atoms in the reference bond. Briefly, we evaluate the change from ε_{ij}° to ε_{ij} in terms of the differences $\Delta q_i = q_i - q_i^{\circ}$ and $\Delta q_j = q_j - q_j^{\circ}$ with respect to the charges in the reference bond. As regards the ε_{ij}° energies, their calculation from Hartree–Fock wave functions involves the use of the virial and Hellmann–Feynman theorems, thus reducing the energy problem to its electrostatic aspects and restricting its solution to molecules in their equilibrium geometry. While feasible in carefully selected, relatively simple, cases, bond energy contributions are in general difficult to obtain at the level of experimental accuracy, using Hartree–Fock data. The good news is that this sort of calculation needs to be done only once for each type of chemical bond, in order to deduce ε_{ij}°. The most interesting part lies ahead, namely the change from ε_{ij}° to ε_{ij} due to the changes in atomic populations at the bond-forming atoms i and j.

The addition of a small amount of electronic charge to an atom in a molecule has two effects. First, a lowering in energy takes place at the atom which has gained electronic charge, much in the same way the energy of an isolated atom decreases as its electron population increases. Second, a change in nuclear–electronic potential energy affects each nucleus "bonded" to the atom whose electron population is modified. These two effects are reflected in the coefficient a_{ij} which expresses by how much the energy contribution of a bonded atom pair ij is affected by a unitary change in net charge at atom i. Similarly, a_{ji} refers to the ij bond and a modification in net charge at atom j. Theoretical expressions are developed on these grounds for the a_{ij} coefficients, and the energy contribution of an ij bond, reflecting now the charge-related effects, becomes

$$\varepsilon_{ij} = \varepsilon_{ij}^{\circ} + a_{ij}\Delta q_i + a_{ji}\Delta q_j.$$

Moreover, the total "bonded" energy contribution is obtained from the sum over all the chemical bonds, i.e.,

$$\sum \varepsilon_{ij} = \sum \varepsilon_{ij}^{\circ} + \sum_i \sum_j a_{ij}\Delta q_i.$$

Finally, including also the small nonbonded interactions, we obtain the following expression for the energy of atomization, ΔE_a^*, of a molecule in its hypothetical vibrationless state at 0 K, i.e.,

$$\Delta E_a^* = \sum \varepsilon_{ij}^{\circ} + \sum_i \sum_j a_{ij}\Delta q_i - E_{nb}^*$$

which is simply the difference in total energy between the isolated atoms from which a given molecule is made up and the molecule itself, excluding, of course, all molecular vibrational energies. The latter equation has been tested for numerous molecules by means of comparisons with experimental results. The calculated atomization energies agree with their experimental counterparts within 0.16 kcal/mol (average deviation). A simple approximation, whose validity can be justified because of the minor role played by the nonbonded interactions, avoids the tedious explicit calculation of the E_{nb}^* terms, with a small loss in precision (~ 0.23 kcal/mol, average deviation), which is still well within experimental uncertainties. The problems encountered with the calculation of the a_{ij} parameters are usually not severe. The main difficulty lies in the accurate evaluation of atomic charges. For this reason, we are presently limited to the study of large classes of homologous compounds for which adequate information concerning charge distributions is presently available. On the other hand, it is certainly pleasing to witness the regularity with which the present theoretical approach reproduces energy differences, even small ones, between closely related compounds—a fact which should encourage us to continue with the study of atomic charges in classes of compounds which have not yet been explored.

At this stage it has become clear that the theory of the chemical bond and molecular energies has developed into a theory of charge density. As discussed above, the basic reason is due to the necessity of restoring molecular electroneutrality, a condition which cannot be ensured by simple bond additivity schemes using a constant energy contribution for each type of chemical bond. Here it is important to note that it is the $\sum_i \sum_j a_{ij}\Delta q_i$ term which ensures the proper charge normalization. Actual molecules, of course, satisfy the electroneutrality requirement simply by allowing all atomic charges to assume their proper values, which are reflections of the appropriate wave function, i.e., ultimately, of the molecular structure. In short, the $\Sigma_i\Sigma_j a_{ij}\Delta q_i$ term is structure-dependent because this is the case with the Δq_i's. Indeed, it represents by far the leading term accounting for the energy differences between isomers because nonbonded Coulomb interactions play an almost negligible role in this respect and $\Sigma\varepsilon_{ij}^{\circ}$ remains constant in iso-desmic structural changes. While, of course, comprehensive calculations of molecular energies require the detailed knowledge of the appropriate charges, an interesting insight can be gained without this knowledge, from an examination of the a_{ij}'s.

To begin with, consider that a negative value for the increment Δq_i in net atomic charge corresponds to an actual increase of electron population at atom i. On the other hand, the a_{ij} coefficients are negative, e.g., $a_{HC} = -0.632$, $a_{CH} = -0.247$, and $a_{CC} = -0.488$ kcal mol^{-1}me^{-1}. Consequently, any increase in electronic charge at the bond-forming atoms leads to larger bond energies, which is a stabilizing effect. So, for example, a gain of 1 me ($\Delta q = -1$ me) at a hydrogen atom stabilizes a CH bond by 0.632 kcal/mol, whereas 1 me added to carbon stabilizes a CH bond by 0.247 and a CC bond by 0.488 kcal/mol. Moreover, taking into account all the bonds formed by a carbon atom, it appears that any electron transfer from hydrogen to carbon results in a stabilization at the molecular level. For example, a transfer of 1 me from a H atom to a C atom engaged in two CH and two CC bonds represents a destabilization of 0.632 kcal/mol at the CH bond and a gain of $(2 \times 0.247) + (2 \times 0.488) = 1.470$ kcal/mol at the CH and CC bonds, for a net gain in stability of 0.838 kcal/mol. In a way, the alkyl hydrogen atoms play the role of a reservoir of electronic charge which, under appropriate circumstances depending on molecular geometry, is called upon to stabilize bonds other than CH bonds, with a net gain in molecular stability. This conclusion holds whenever the sum $\Sigma_j a_{ij}$, measuring the stabilization of all the bonds formed by atom i by an electronic charge added to it, is more negative than a_{HC}. This is the case for the ether and carbonyl oxygen atoms, with $\Sigma_j a_{ij} \simeq -1.00$ and -0.665 ± 0.01 kcal mol^{-1}me^{-1}, respectively. The following examples illustrate the role of the hydrogen atoms acting as "electron reservoirs". Twistane isomerizes into adamantane with a loss of electronic charge at the hydrogen atoms in favor of the carbon atoms (H $\rightarrow$ C transfer). This electron enrichment at the carbon skeleton causes adamantane to be the thermodynamically more stable isomer. On the other hand, the stabilization accompanying isomerizations like $C_2H_5CHO \rightarrow CH_3COCH_3$ and CH_3CO-n-$C_3H_7 \rightarrow CH_3CO$-$i$-$C_3H_7$ is largely traced back to an electron enrichment at the oxygen atom at the expense of electronic charge at the hydrogen atoms (H $\rightarrow$ O transfer). Similarly, significant stabilizations accompanying structural changes involving ether α-carbons, like CH_3O-n-$C_3H_7 \rightarrow CH_3O$-i-C_3H_7, are mainly due to H $\rightarrow$ O charge transfers. However, a structural modification affecting ether β-carbons, as in CH_3O-n-$C_4H_9 \rightarrow CH_3O$-i-C_4H_9, results in a prevailing H $\rightarrow$ C transfer and, accordingly, in a stabilization of the alkyl part, just as is the case in paraffins.

These examples illustrate a way of discriminating between a stabilization occurring mainly "within" the alkyl part and a stabilization involving a functional group.

It is certainly pleasing that this insight into the structure-related fine-tuning of molecular energies by local charges derives exclusively from basic and clearly identified physical concepts, namely (*i*) the Politzer–Parr partitioning of molecular energies in terms of "atomic-like contributions"[7] which emphasizes the dependence of atomic and molecular energies upon the electrostatic potentials at the nuclei, (*ii*) the relationships[8] between nuclear–nuclear, nuclear–electronic, electronic–electronic and Hartree–Fock orbital energy components, and the total energy, of which Politzer's approximations[9] are an important part and, finally, (*iii*) the reduction of the problem of molecular energies to its electrostatic aspects[10], featuring the important role of local charges. However, while conceptually simple, this approach does not leave much room for errors, precisely because of the absence of postulated effects which could be parametrized in empirical fashion. Applications are not only a matter of computer and financing technology, but also of dedication to accuracy. At this level, "small" errors (namely, in the evaluation of charges) easily translate into serious discrepancies. The agreement between theory and experiment demonstrated here for a body of over 100 molecules is certainly encouraging, also as regards the various aspects related to the evaluation of atomic charges. Finally, while this type of theory refers only to molecules in their hypothetical vibrationless state, appropriate comparisons with enthalpy data offer a new access to numerous data for molecular vibrational energies, which are certainly worth additional investigations.

REFERENCES

1. V.W. Laurie and J.S. Muenter, *J. Am. Chem. Soc.*, **88**, 2883 (1966).
2. H. Henry and S. Fliszár, *Can. J. Chem.*, **52**, 3799 (1974).
3. R.W. Taft and J.C. Lewis, *J. Am. Chem. Soc.*, **80**, 2436 (1958); *Tetrahedron*, **5**, 210 (1959); S.K. Dayal, S. Ehrenson, and R.W. Taft, *J. Am. Chem. Soc.*, **94**, 9113 (1972).
4. L.S. Levitt and H.F. Widing, *Prog. Phys. Org. Chem.*, R.W. Taft, Ed., **12**, 119 (1976).
5. W.J. Hehre, R.W. Taft, and R.D. Topsom, *Prog. Phys. Org. Chem.*, R.W. Taft, Ed., **12**, 159 (1976).
6. R.F.W. Bader, S.G. Anderson, and A.J. Duke, *J. Am. Chem. Soc.*, **101**, 1389 (1979); K.B. Wiberg, *J. Am. Chem. Soc.*, **102**, 1229 (1980).
7. P. Politzer and R.G. Parr, *J. Chem. Phys.*, **61**, 4258 (1974).
8. S. Fliszár, M. Foucrault, M.-T. Béraldin, and J. Bridet, *Can. J. Chem.*, **59**, 1074 (1981).
9. P. Politzer, *J. Chem. Phys.*, **64**, 4239 (1976); P. Politzer, *J. Chem. Phys.*, **70**, 1067 (1979); T. Anno, *J. Chem. Phys.*, **72**, 782 (1980).
10. S. Fliszár, *J. Am. Chem. Soc.*, **102**, 6946 (1980).

Summary of Final Equations and Input Parameters

The final equations and input parameters given here are intended to serve as a "guide-to-the-user".

1. The energy of atomization of a nonlinear molecule in its hypothetical vibrationless state, ΔE_a^*, can be deduced from experimental data by means of Eq. 5.5 (Chapter 5.2),

$$\Delta E_a^* = \sum_i n_i [\Delta H_f^\circ(A_i) - \frac{5}{2} RT] + \text{ZPE} + (H_T\text{-}H_0) - \Delta H_f^\circ$$

in which ΔH_f° is the standard enthalpy of formation (gas, 298.15 K) of the molecule and $T = 298.15$ K. The heat content, $H_T\text{-}H_0$, is tabulated for numerous molecules (e.g., in Ref. 1); alternatively, it can be deduced from fundamental vibrational frequencies using Einstein's function (Eq. 5.7) for E_{vibr} and the equation $H_T\text{-}H_0 = F_{\text{vibr}} + 4\ RT$. In the harmonic oscillator approximation, the zero-point energy ZPE is as indicated in Eq. 5.6. Convenient approximations for ZPE + $H_T\text{-}H_0$ are given in Chapter 7, namely, Eq. 7.2 (noncyclic saturated hydrocarbons), Eq. 7.5 (cycloalkanes containing m six-membered rings), Eq. 7.6 (ethylenic hydrocarbons), and Eq. 7.7 (dialkylethers). The standard enthalpies of formation of the gaseous atoms A_i (whose number in the molecule is n_i) are taken from Ref. 2, namely, $\Delta H_f^\circ(C) = 170.89$, $\Delta H_f^\circ(H) = 52.09$, and $\Delta H_f^\circ(O) = 59.54$ kcal/at.g.

2. Saturated hydrocarbons containing m ($\geqslant 0$) six-membered (chair and/or boat) rings can be calculated in a bond-by-bond approach by means of Eq. 5.46. Using the parameters indicated in Tables 5.6 and 5.7, Eq. 5.46 gives the bond energy contributions

$$\varepsilon_{CH} = 106.806 - 0.247\Delta q_C - 0.632\Delta q_H \text{ kcal mol}^{-1}$$

$$\varepsilon_{CC} = 69.633 - 0.488\Delta q_{Ci} - 0.488\Delta q_{Cj} \text{ kcal mol}^{-1}$$

with Δq_C and Δq_H expressed in 10^{-3} electron ($=$ me) units. The Δq_C's are readily deduced from ^{13}C NMR shifts relative to ethane (taken at 5.8 ppm from TMS) using Eq. 4.10, i.e.,

$$\Delta q_C = -0.148\delta_C \text{ (me)}.$$

For deducing the Δq_H's, we calculate, first, the net charge of each carbon from $q_C = 35.1 + \Delta q_C$ (me) and the sum Σq_C. The average net charge of one H atom is $-\Sigma q_C$ divided by the number (n_H) of hydrogen atoms in the molecule and, hence, Δq_H(average) $= -\Sigma q_C/n_H + 11.7$ me for use in calculations of ε_{CH} (examples are given in Chapter 6.2). The sum of all ε_{CH}'s and ε_{CC}'s over the molecule yields the "bonded part" of ΔE_a^*, i.e. $\Delta E_a^{*\text{bonds}}$. Straightforward calculations of $\Delta E_a^{*\text{bonds}}$ for $C_nH_{2n+2-2m}$ hydrocarbons are, however, more conveniently performed by means of the equation

$$\Delta E_a^{*\text{bonds}} = 80.723(n + m - 1) + 104.958(2n + 2 - 2m)$$
$$+ 0.0356\sum N_{CC}\delta_C + 0.0529\sum \delta_C \text{ kcal mol}^{-1}$$

(from Eqs. 6.3, 6.4 and 6.7, Chapter 6), where N_{CC} is the number of CC bonds formed by the carbon whose shift relative to ethane is δ_C and $(n + m - 1)$ and $(2n + 2 - 2m)$ are, respectively, the numbers of CC and CH bonds in the molecule. In order to obtain $\Delta E_a^* = \Delta E_a^{*\text{bonds}} - E_{nb}^*$ (Eq. 5.29), we still have to evaluate E_{nb}^* from Eq. 5.30. This tedious calculation can be avoided by means of the following approximation for ΔE_a^* (Eq. 6.11),

$$\Delta E_a^* \simeq 710.54(1 - m) + 290.812(n - 2 + 2m) + 0.03244\sum N_{CC}\delta_C$$
$$+ 0.05728\sum \delta_C \text{ kcal mol}^{-1}$$

which involves only a small loss in precision (average deviation $= 0.22$ kcal mol^{-1}, Chapter 6.7). For obtaining ΔH_f° (gas, 298.15 K), one proceeds as indicated in the previous section and uses Eq. 7.2 or 7.5 for the numerical estimate of ZPE $+ H_T$-H_0. It follows (Chapter 7.2) that

$$\Delta H_f^\circ = \text{ZPE} + H_T\text{-}H_0 - 27.692(1 - m) - 20.184n - 0.03244\sum N_{CC}\delta_C$$
$$- 0.05728\sum \delta_C \text{ kcal mol}^{-1}.$$

Calculations conducted along these lines are accurate within thermochemical experimental uncertainties, except in cases exhibiting an unusual steric crowding (as with two *tert*-butyl groups attached to a same carbon). The suspected reasons originate possibly in a breakdown of the simple scheme for evaluating ZPE $+ H_T$-H_0 energies and/or in a failure of the charge–NMR shift relationship due to slight changes in hybridization (Chapter 4.7). Additional spectroscopic work is highly desirable in this area.

3. Bond-by-bond calculations of simple alkenes C_nH_{2n} require the ε_{CH} and ε_{CC} energies described in Section 2 for the bonds involving sp^3 carbons. In addition, one uses (Chapter 8.4)

$$\varepsilon_{C=C}(sp^2-sp^2) = 139.27 - 0.183\Delta q_{Ci} - 0.183\Delta q_{Cj} \text{ kcal mol}^{-1}$$

where the Δq_C's (in me) are taken with respect to the net charge of the ethylene carbon atom (7.7 me). The other bonds formed by the sp^2 carbons (identified as Cα) are

$$\varepsilon_{CH}(sp^2-H) = 110.68 + 0.454\Delta q_{C\alpha} - 0.632\Delta q_H \text{ kcal mol}^{-1},$$

$$\varepsilon_{CC}(sp^2-sp^3) = 77.69 + 0.275\Delta q_{C\alpha} - 0.488\Delta q_C \text{ kcal mol}^{-1}.$$

The $\Delta q_{C\alpha}$'s (relative to the carbon of ethylene) are deduced from an approximation based on the ^{13}C NMR shifts relative to ethylene (Chapters 8.2.2, 8.2.3), namely, $\Delta q_{C\alpha} \simeq 0.15\delta_C$ (me) whereas Δq_C (for sp^3 carbon) is, as usual, $-0.148\delta_C$ (me). Note that for the sp^2 carbons $q_C \simeq 7.7 + 0.15\delta_{C\alpha}$ (me). The Δq_H's are obtained from charge normalization, as explained above in Section 2 (see Chapter 8.4 for numerical examples). Calculations of ΔE_a^* are carried out more efficiently by means of the following general formulas. Monoalkyl- and *trans*-dialkylsubstituted ethylenes, as well as tetramethylethylene, are best treated by the approximation (Eq. 8.12, Chapter 8.2)

$$\Delta E_a^* = \sum \varepsilon_{ij}^\circ + 0.0356\sum N_{CC}\delta_{C\neq\alpha} + 0.0529\sum\delta_{C\neq\alpha} + 0.18\sum\delta_{C\alpha}$$
$$+ 7.393n + 4.19\sum N_{C\alpha C} - 19.06 \text{ kcal mol}^{-1}$$

whereas *cis*-disubstituted and trisubstituted ethylenes are described by

$$\Delta E_a^* = \sum \varepsilon_{ij}^\circ + 0.0356\sum N_{CC}\delta_{C\neq\alpha} + 0.0529\sum\delta_{C\neq\alpha} + 0.18\sum\delta_{C\alpha}$$
$$+ 7.393n + 4.0\sum N_{C\alpha C} - 18.12 \text{ kcal mol}^{-1}$$

as indicated in Chapter 8.3. The reference bond energies ε_{ij}° are 69.633 for *any* CC simple bond, 106.806 for *any* CH bond, and 139.27 kcal mol^{-1} for the CC double bond (from Table 5.6). The $\delta_{C\neq\alpha}$ chemical shifts of sp^3 carbons are taken relative to ethane and the $\delta_{C\alpha}$ shifts of sp^2 carbons relative to ethylene. The $\Sigma N_{C\alpha C}$ term represents the total number of CC simple bonds formed by the *two sp²* carbon atoms and n is the total number of C atoms. The standard enthalpies of formation ΔH_f° (gas, 298.15) are now easily deduced as outlined in Section 1, using the appropriate expression (Eq. 7.6, Chapter 7.5) for ZPE + H_T-H_0. These calculations should not be applied in situations of extreme steric crowding (e.g., when two *tert*-butyl groups are attached to the same C atom).

4. Atomization energies of dialkylethers are conveniently deduced from Eq. 9.6. With the appropriate parameters (Chapter 9.3) and the a_{ij}'s of Table 5.7 (1 a.u. = 627.51 kcal mol^{-1}), this equation becomes

$$\Delta E_a^{*\text{bonds}} = \sum \varepsilon_{ij}^\circ + 0.0356\sum N_{CC}\delta_C + 0.0529\sum\delta_{C\neq\alpha} + 0.1217\sum\delta_{C\alpha} + \lambda_4\delta_O$$
$$+ 7.393n + 0.92\sum N_{CC}^\alpha - 10.67 \text{ kcal mol}^{-1}.$$

The standard bond contributions ε_{ij}° are (from Table 5.6) 69.633 for *any* CC, 106.806 for *any* CH, and 79.78 kcal mol^{-1} for one CO bond. The $\delta_{C\neq\alpha}$ shifts are relative to ethane (at 5.8 ppm from TMS) and the $\delta_{C\alpha}$ shifts are relative to the α-carbon of diethylether. The $\Sigma N_{CC}\delta_C$ term includes both $\delta_{C\alpha}$ and $\delta_{C\neq\alpha}$. The ^{17}O NMR shifts are relative to that of diethylether. The parameter n is the number of C atoms and ΣN_{CC}^{α} is the number of CC bonds formed by the two α-carbons. Contrasting with most a_{ij}'s which can be taken as constants because the $(1/2)(\partial^2 E/\partial N^2)\Delta q$ term in Eq. 5.47 is negligible in comparison with $(\partial E/\partial N)$, a_{OC} and, hence, λ_4 (Eq. 9.8), depend on Δq_O. In practice, however, it is sufficiently accurate to take $\lambda_4 = 0.1007$ when $\delta_O < 0$ and $\lambda_4 = 0.0994$ kcal mol^{-1} ppm^{-1} when $\delta_O > 0$ (Chapter 9.3). A detailed example of bond-by-bond calculations is worked out in Chapter 9.3 and may serve as a guide in this sort of evaluation of ΔE_a^{*bonds}. Inclusion of nonbonded energies gives $\Delta E_a^* = \Delta E_a^{*bonds} - E_{nb}^*$. Rough evaluations of E_{nb}^* can be made on the basis of the results given in Table 9.2: for normal alkyl groups larger than C_2H_5, E_{nb}^* becomes more negative by ~ 0.13 kcal mol^{-1} for each added CH_2 group (see also p. 111) but branching (e.g., for iso-C_3H_7 instead of n-C_3H_7) makes E_{nb}^* more positive by ~ 0.08 kcal mol^{-1}. The appropriate ZPE + H_T-H_0 energies can be deduced from Eq. 7.7 (Chapter 7.7) and, finally, ΔH_f° (gas, 298.15) can be obtained from the equation given in Section 1. Acetals should not be calculated in this manner because their charge–^{17}O NMR shift relationship differs from that of the simple ethers.

5. The energy formula describing noncyclic carbonyl compounds is (from Eq. 9.9, Chapters 9.2 and 9.4)

$$\Delta E_a^{*bonds} = \sum \varepsilon_{ij}^\circ + 0.0356 \sum N_{CC}\delta_C + 0.0529 \sum \delta_{C\neq\alpha} + 0.0675\delta_{C\alpha} + \lambda_4'\delta_O$$
$$+ 7.393n + 5.07N_{CC}^\alpha - 19.42 \text{ kcal mol}^{-1}.$$

The ε_{ij}°'s are (Table 5.6) $\varepsilon_{CH}^\circ = 106.806$, $\varepsilon_{CC}^\circ = 69.633$ and $\varepsilon_{CO}^\circ = 179.40$ kcal mol^{-1}. The $\delta_{C\neq\alpha}$ and $\delta_{C\alpha}$ (for the carbonyl-C atom) ^{13}C shifts are relative to ethane and acetone, respectively, while δ_O is relative to the ^{17}O NMR shift of acetone. The parameter n is the total number of C atoms and N_{CC}^α is the number of CC bonds formed by the carbonyl-C atom. The $\Sigma N_{CC}\delta_C$ term includes both $\delta_{C\alpha}$ and $\delta_{C\neq\alpha}$. The λ_4' parameter (Eq. 9.11) depends strongly on Δq_O (Table 9.5) because this is the case with a_{OC} which must be recalculated for each compound under study, using Eq. 5.47. From Table 5.5 and 5.7, as well as Δq_O (carbonyl) $\simeq 2.7\delta_O$ me (Eq. 9.12, Chapter 9.4), it follows that $a_{OC} = -1.065 - 6.8 \times 10^{-4} \times \delta_O$ a.u. and $\lambda_4' = -(0.058 + 6.8 \times 10^{-4}\delta_O)$ a.u., i.e.,

$$\lambda_4' = -(0.098 + 1.15 \times 10^{-3} \times \delta_O) \text{ kcal mol}^{-1} \text{ ppm}^{-1}.$$

Numerical examples for this type of calculations are indicated in Tables 9.4 and 9.5. Nonbonded energies required for deducing $\Delta E_a^* = \Delta E_a^{*bonds} - E_{nb}^*$ are indicated in Table 9.6. For larger normal alkyl groups, E_{nb}^* becomes more negative by ~ 0.13 kcal mol^{-1} for one added CH_2, but branching is expected to make E_{nb}^* more positive by ~ 0.08 kcal mol^{-1}. The ΔH_f° (gas, 298.15) values are deduced as indicated in Section 1 with the help of ZPE + H_T-H_0 constructed from the approximations described in Chapter 7.6.

REFERENCES

1. F.D. Rossini, "Selected Values of Physical and Thermodynamic Properties of Hydrocarbons and Related Compounds", Carnegie Press, Pittsburgh, PA, 1952.
2. D.R. Stull and G.C. Sinke, *Adv. Chem. Ser.*, **18** (1956).

Author Index

Subject Index